Earned Intelligence

$$I_{earned} = M_{meaning} \times S_{systems} \times B_{behavior}$$

Deepak Rana

Dedication

To my mother and father, whose quiet sacrifices and unspoken lessons became the soil in which my dreams could take root.

To my wife, whose love steadies me and whose faith in me has been the light I return to, again and again.

To my son, whose very existence reminds me of what is possible, and whose laughter gives meaning to every effort.

I carry within me the voices of family and friends who, in ways both gentle and demanding, shaped how I see the world and how I tell its stories. Their presence has been a mirror, a challenge, and a gift.

And above all, I bow in gratitude to God—my compass through uncertainty, my shelter in storms, and the quiet strength behind every page of this book.

This work is not mine alone. It is the sum of love, faith, and countless moments of grace.

With all my heart—thank you.

CONTENTS

Editor's note

This book captures a truth witnessed across countless enterprises: technology rarely fails for lack of talent—it fails for lack of shared understanding.

The story of DWL, though fictionalized for flow, reflects challenges that are universally real.
Fragmented systems.
Semantic drift.
Ethical fatigue.

The slow, relentless climb toward data maturity. These are not the problems of one company; they are the defining tests of modern enterprise life. The characters in these pages represent every leader, engineer, and analyst who has wrestled with the gap between what systems know and what organizations truly understand. Their conflicts are drawn from boardrooms and data centers alike, where meaning too often dissolves into metrics.

The lessons that emerged are practical and urgent:
- Define language before deploying models.
- Govern data before trusting predictions.
- Pair ethics with automation so that speed never outruns conscience.

DWL's transformation proves that maturity is not measured in terabytes but in trust. Their journey marks the transition from artificial intelligence to Earned Intelligence[1]—transparent, explainable, and accountable.

What DWL built can be replicated, and the framework they developed offers a roadmap for any organization striving to make intelligence sustainable.

This is not a manual for machines. It is a guide for cultivating intelligence that can be trusted—by executives, by engineers, and by every person who must decide what truth means before technology defines it for them.

If it helps leaders think before they predict, its purpose has been fulfilled

---

[1] **Earned Intelligence**—The ability to make decisions from data that is governed, models that are explainable, and outcomes that are auditable—measured monthly against defined trust SLAs.

DEEPAK RANA

Preface

The Moment Truth Became Work

The greatest threat to artificial intelligence is not bias, ethics, or even flawed data—it is language.

Most organizations never recognize the danger until it is too late. Reports contradict one another, dashboards disagree, and teams argue not about outcomes but about what "success" even means. Psychologists call it **semantic drift**—when the same word carries different meanings for different people. This erosion of shared meaning undermines trust long before any algorithm fails. What appears to be technical collapse is, in truth, human confusion disguised as precision.

At DWL, the fracture revealed itself in silence. Three dashboards. One number. Three conflicting truths. No one blamed the system. No one raised their voice. They simply stared, realizing that intelligence had outpaced understanding.

Daniel Shaw, the CEO, broke the silence with a sentence that would define the company's transformation:

"We built intelligence, but not agreement."

That moment marked the turning point. What followed was not glamorous innovation but the unrelenting labor of alignment.
Definitions became debates.
Debates became reform.
Out of that struggle emerged **The Language of Truth**—a recognition that before machines could predict the future, humans had to agree on the present. The process was grueling. Arguments over single words exposed fractures in how people thought about the business. Midnight reconciliations blurred into days. The lesson was humbling: trust cannot be automated. Yet as language aligned, data stabilized. As definitions hardened, performance improved.

Two years later, DWL did not simply run AI—it understood it. The company had built what many promise but few achieve: **Earned Intelligence.**

This book tells that story. It is fiction only in form—because the experiences behind it are real. Every conflict, every lesson, every resolution is drawn from lived experience—observed, reframed, and retold. It is the story of professionals who discovered that clarity is not a byproduct of intelligence;

it is its foundation. Before prediction comes comprehension. Before automation comes agreement. Because meaning, not math, is what makes intelligence worth trusting.

Prologue

The shadow protocol

Kara M. Sloane preferred arriving while the building was still half-asleep.

At 6:30 a.m., the halls at DWL's global headquarters were quiet enough to think, and the systems hadn't yet absorbed a full day of human interpretation.

For a senior data quality analyst, that quiet mattered. She logged in, opened the overnight reconciliation reports, and did what she always did—confirmed that yesterday still agreed with today. Most mornings, it did.

This morning, one metric didn't.
A variance of 0.8%.
Tiny.
Almost decorative—and wrong.

Kara ran the usual checks—no upstream changes, no pipeline[2] failures, no schema[3] edits, no anomalous row counts. The data was perfect in every mechanical way.

The meaning wasn't.

She compared it against the past week. Same inputs, same logic, different outcome. Someone, somewhere, had reinterpreted the metric—without a change ticket, without communication, without alignment.
It wasn't an error.
It was behavior.

She opened a message to Priya Nayar, Head of Governance.

Kara:
One asset metric shifted by 0.8%. Feeds clean. Possible definition drift.

Priya responded eight minutes later—not with a message, but by walking into the analytics room. That alone told Kara everything. Priya didn't waste steps.

"Show me," Priya said.

---

[2] **Pipeline**—A structured sequence of automated steps for preparing data, training models, and deploying results. for example, Pipelines in Azure ML and AWS SageMaker processed customer order data end-to-end to deliver backlog forecasts
[3] **Schema**—A formal structure that defines how data is organized, labeled, and related within a system. for example, Schemas in Azure SQL Database and AWS RDS standardized asset records so customer order data could be integrated consistently across systems

Kara pulled up the dashboard. "See here—yesterday's result. Now this mornings'. Identical inputs, different outcome."

"Who owns the metric?" Priya asked.

"That's the problem," Kara said. "Operations claims it. Finance adjusts it. Sales uses a customized version."

Priya exhaled slowly. "So: three owners, three interpretations."

"And no agreement," Kara added.

Priya nodded. "This isn't a system issue. This is the pattern behind the system."

She glanced at the time—7:04 a.m. "We have the quarterly board review tomorrow. This variance will show up somewhere in their numbers."

"You think they'll notice 0.8%?" Kara asked.

"No," Priya said. "They'll notice the disagreement."

Kara hesitated. "Do we fix it?"

"If we fix it," Priya said, "we make the surface look clean. If we leave it, they'll see the drift for what it is." She paused. "We need Daniel."

Daniel Shaw, DWL's CEO, arrived a few minutes later. He listened, quietly, through the whole explanation—variance, drift, ownership conflict.

Then he asked the one question Kara hoped he wouldn't.
"Is there a single, documented definition for this metric?"

Kara shook her head. "No. Everyone uses their own version."

Daniel looked at both of them. "Then the number isn't broken. Our alignment is."

He studied the screen again. "If we correct this before tomorrow, the board will see stable charts and assume stability. They need to feel the inconsistency."

"You want to expose it," Priya said.

"Yes," Daniel replied. "If the board doesn't experience the friction, they won't understand the root."

Kara shifted in her seat. "It goes against every instinct not to fix something."

"I know," Daniel said gently. "But this isn't about repair. It's about visibility."

He stepped back. "Tomorrow, three teams will report three versions of truth. A small shift like this will amplify the divide."

Priya folded her arms. "We've seen it before—EPAC[4] behavior."

Kara glanced up. "EPAC?"

Priya answered. "Emergent Pattern of Administrative Contradictions.

EPAC—wasn't an organized group.
It wasn't a department.
It wasn't even intentional.
It was a behavior.
A drift.
A quiet accumulation of shortcuts and "temporary fixes" that had become permanent. It's how small interpretive changes accumulate until no two teams speak the same operational language."

Daniel nodded. "Not malicious. Just ungoverned."

Kara looked again at the 0.8% variance. "So, what should I do?"

"Document everything," Daniel said. "Do not correct it. Track any further drift."

Kara nodded. "Understood."

Daniel turned toward Priya. "After tomorrow, we begin addressing this. Properly."

Priya asked, "You already know who needs to help lead it?"

---

[4] **Emergent Pattern of Administrative Contradictions (EPAC)**—was not a group but a behavior. But those behaviors tended to originate from the same clusters of roles under pressure to deliver numbers, fast.
DEEPAK RANA

Daniel nodded. "Yes. But he won't understand the scope until he sees it himself."

Kara wasn't sure who he meant, but she knew Daniel rarely answered questions prematurely.

"Tomorrow," Daniel said, "is the beginning."

When they left, the room became quiet again.
Kara stared at the screen—two points, 0.8% apart.
It didn't look like a crisis.
But it felt like the start of one.
She closed her laptop and prepared to move on with her morning checks. The variance vanished from the screen but not from what the next day would reveal.

PART I—THE AWAKENING

When DWL realizes its challenge isn't data or technology, but meaning itself. Four chapters reveal how misalignment forms, grows, and finally becomes undeniable.

CHAPTER 1—The Breaking Point

Three versions of revenue, three executives—limited trust.

The quarterly board review began ten minutes late, which irritated Daniel Shaw more than anyone else in the room. He valued time almost as much as he valued clarity, and this morning promised to test both.

The boardroom overlooked the city—glass on three sides, a paneled wall on the fourth. A symbol of transparency, Daniel often joked, except on days when the numbers didn't agree.

Today was one of those days.

The leadership team took their seats.

Elena Park, CFO, disciplined and meticulous, carried a binder of reconciled figures she had personally reviewed twice.

Aaron Cole, COO, the operational backbone of DWL, who believed numbers were only useful if they helped plants run better.

Maya Chen, Chief Commercial Officer, sharp and grounded in customer reality.

Priya Nayar, Head of Governance, quiet but precise, whose influence came from the consistency of her questions rather than the volume of her voice.

At the support table along the side sat Ethan Anderson, Senior Manager in Strategic Analytics. Not yet inner circle, but increasingly unavoidable. He had a way of noticing things others stepped over.

The main screen displayed:
DWL—Q2 Performance Review

No one knew about the 0.8% drift Kara had found. But they were about to feel its consequences.

Daniel opened with his usual brevity. "Let's begin."

He nodded to Elena.

Elena stepped to the front. "Finance closes Q2 at a margin of 18.4%," she said. "Up 1.1 points from Q1."
Charts appeared—clean, logical, defensible.

When she finished, Aaron stood.
"Operations closes Q2 at a margin of 16.9%," he said.

The room shifted. Quietly. But noticeably.

Then Maya added, "Commercial performance reflects closer to 15.8% due to strategic concessions."

Now the gap was nearly three points.
Three leaders.
Three truths.
One company.

A board member finally asked, "Which of you is correct?"

All three were.
That was the problem.

Daniel waited until the tension peaked.
Then he called on Ethan.
"Can you explain why we're seeing three different margins?"

Ethan moved to the table with practiced caution.
"The issue isn't data quality," he said. "It's definitional. Each function uses the word 'margin' differently. Finance uses full P&L. Operations uses controllable plant costs. Commercial adjusts for concessions. All valid. None aligned."

A director asked, "Is this new?"

"No," Ethan said. "What's new is the visibility."

Daniel watched the board absorb this.
"And if we continue like this?" Daniel asked him.

"If we continue like this," Ethan said softly, "we'll spend more time reconciling the past than planning the future. And any analytics or artificial intelligence (AI)[5] we consider later will inherit and amplify these contradictions."

The phrase hung in the air.
Not a threat. A truth.

---

[5] **Artificial Intelligence (AI)**—A technology that enables systems to automatically process information, recognize patterns, and generate insights or actions.

"This," Daniel said, looking around the room, "is our breaking point."

He turned to the board. "Before we invest another dollar in analytics or AI, we must align meaning. We must stop confusing interpretation with insight."

He looked at his team.

"Elena. Aaron. Maya. Priya. Ethan. We will define the core terms together. Starting now."

"What do you call this effort?" a board member asked.

Daniel answered without hesitation.

"The Language of Truth."

And with that, the next 18 months—and the 6-month stabilization that would follow—quietly began.

Reflection—What Broke and What Begins
What we observed
Three leaders, three valid numbers, one fractured reality.

What it means
Without shared definitions, truth becomes negotiable—and intelligence becomes untrustworthy.

What will happen next
Pause AI.
Align meaning.
Begin The Language of Truth initiative.

Aphorism
Intelligence without agreement is precision without trust.

Guiding Principle
Shared meaning precedes shared measurement.

The Law of Semantic Alignment
When definitions drift, truth fragments, and all analytics become unreliable.

Equation
$$Q_{decision} = K_1 \times A_{alignment}$$

Technical Explanation
Decision quality $Q_{decision}$ improves as semantic alignment $A_{alignment}$ improves. If alignment is low, adding more dashboards, models, or data science will not materially improve decisions.

Symbols
$Q_{decision}$: Overall quality of decisions.
$A_{alignment}$: Degree of shared definitions across teams.
K_1: Positive scaling constant that converts alignment into decision quality units.

CHAPTER 2—Ethan's Awakening

Introspection

Ethan Anderson spent the morning after the board meeting replaying the questions in his mind—not the board's questions, but Daniel's.

"Has anyone documented which definition is 'official' for this metric?"

He had heard those words before—variations of them—in reconciliations across three years at DWL, each time revealing inconsistencies everyone worked around but no one solved. But yesterday had been different. Yesterday, the misalignment stood under a spotlight the size of the boardroom.

Today, Ethan felt something shift—not dramatic, not cinematic, but unmistakable.

Someone in leadership had finally said out loud what he had only hinted at in his analyses:
The company wasn't disagreeing on performance.
It was disagreeing on language.

And somehow, he was now pulled into the center of it.

The Quiet After the Board Meeting

Ethan arrived early the next morning, earlier than usual. Not Kara-early—no one arrived as early as Kara—but early enough to feel like he had a head start on the day. The office was quiet, lights flickering on automatically as he passed the sensors. He walked straight to the analytics lab.

Kara stood there, as if she had been waiting—not for him, but for the day to begin.

"Morning," Ethan said.

"Morning," she replied without turning, focused on her screen. "You're early."

"Couldn't sleep," he admitted.

Kara didn't smile, but something in her expression softened. "Yesterday will do that to you."

He hesitated. "Did you expect Daniel to call me out like that?"

"No," she said. "But I wasn't surprised."

"Why?"

"Because you don't pretend things align when they don't."
She tapped the monitor. "That's rarer than it should be."

Before he could reply, Priya appeared behind him.

"Both of you," she said, "conference room in five minutes."
She didn't wait for acknowledgment. She walked past them, calm, composed, already moving ahead of the moment.

Kara closed her laptop. "Welcome to your induction."

"Induction into what?" Ethan asked.

Kara raised an eyebrow. "Into seeing the real work."

The Awakening—The Meeting with Priya

Priya was not a dramatic person. If anything, she was the opposite: measured, quiet, rarely using more words than necessary. Yet there was something undeniably intense about her presence—the kind of intensity that doesn't demand attention, but earns it.

She stood at the head of the small room; a single slide displayed on the monitor:

Term Alignment—Assessing the First Ten

Ten terms. Ethan recognized them immediately:
1. Margin
2. Active Customer
3. Forecast Accuracy
4. Backlog
5. On-Time Shipment
6. Asset Utilization
7. Order
8. Exception
9. Priority
10. Fulfilled

"These," Priya said as they entered, "are the starting point."

Ethan sat. Kara took the seat beside him.

Priya continued, "The board approved Daniel's proposal, but the approval is the easy part. Alignment is the work."

She turned to Ethan. "Yesterday, you said something important: that we spend more time reconciling the past than planning the future. That is precisely the problem we're solving."

Ethan felt a mix of pride and unease. "So, what do you need from me?"

"Not more dashboards," Priya said. "Not more reports. What I need is clarity—your clarity. You see inconsistencies others rationalize away."

Kara added, "He sees them because he traces definitions, not just numbers."

"That's exactly why you're here," Priya said.
She clicked to the next slide.

Term	Current State	Degree of Drift
Margin	3 definitions	High
Active Customer	4 interpretations	High
Order	6 primary versions	High
Backlog	3 partial ownerships	High
On-Time Shipment	1 term, 5 rules	Severe
Asset Utilization	2 versions	Medium
Exception	6 local meanings	Severe
Priority	4 tiers, inconsistent	High
Fulfilled	2 competing views	Medium
Forecast Accuracy	3 formulas	High

Figure I—Ten Terms

Ethan stared at the grid. He knew some of this, but seeing it quantified changed everything.

"This," Priya said, "is the matrix."

Kara stiffened slightly. Priya's use of the word was deliberate.

"Not a simulation. Not some mythology. A matrix of meanings—overlapping, contradictory, shifting without governance."

She turned to Ethan. "And you, whether you realize it or not, have been inside it longer than you think."

Ethan's First Test

Priya handed him a printout. "This is yesterday's reconciliation. Walk me through it."

Ethan skimmed the sheet. He recognized the patterns immediately:
- manual overrides
- adjusted definitions
- logic changes that lived only in someone's spreadsheet
- undocumented assumptions
- small departmental "fixes"
- EPAC fingerprints everywhere

At the bottom of the page was the number Kara had discovered:
Variance: 0.8% (unexplained)

Ethan circled it with his thumb. "The logic didn't change. The meaning did."

Priya nodded. "Explain."

So, he did.
He traced how Finance adjusted the metric for corporate allocations.
How Operations ignored those allocations entirely.
How Sales applied customer-level concessions after the fact.
How each team believed they were right.
How none of them were wrong.
And how all of them produced truths that could not coexist.

When he finished, the room was silent.

Kara whispered, "I told you he sees it."

Priya folded her hands. "Ethan, this is not about intelligence systems. Not yet. This is about the conditions under which intelligence fails."

She walked around the table and stood behind him—not intimidating, but grounding.
"Your job is not to fix the metric. Your job is to show us the map."

"The map?" Ethan asked.

"The map of drift. The map of how meaning moves when no one governs it." She paused. "The map of the matrix."

Daniel's Invitation

Later that afternoon, Ethan was called to Daniel's office.

Daniel's assistant simply said, "He's expecting you."

Ethan entered. Daniel stood by the window, hands clasped behind him, looking at the city.
"You handled yourself well yesterday," Daniel said without turning.

"Thank you," Ethan replied.

Daniel faced him. "You didn't overstate the problem. You didn't minimize it. You described it exactly as it is."

Ethan stayed quiet.

Daniel continued, "The next 18 months will define the next decade of this company. If we fail at this, nothing else we build will matter—not automation, not AI, not transformation."

Ethan nodded. "I understand."

Daniel looked at him closely. "No, not yet. But you will."

There was something in Daniel's tone—an invitation, a warning, and a responsibility, all at once.

"What do you need from me?" Ethan asked.

Daniel walked to his desk, pulled out a folder, and handed it to him.
"This is the initial charter for The Language of Truth. It's not finalized. I want your input."

Ethan hesitated. "You want me to?"

"Yes," Daniel said simply. "Because you see the fractures. And because you don't accept them as normal."

Ethan opened the folder. It wasn't complicated. It was brutally simple:
— Ten terms
— Clear owners
— Governance structure
— Approval rules
— Version control

- — Change logs
- — Cross-functional sign-offs

But the simplicity felt deceptive, like a doorway hiding a much larger room.

"This is the first step," Daniel said. "Before we touch any system, any automation, or any Artificial Intelligence (AI) model—we redefine the language of the company."

He paused.
"And you will help us build it."

Ethan felt the weight of it—not pressure, but clarity.
"I'm in," he said.

Daniel nodded. "Good."

As Ethan turned to leave, Daniel added one more thing.
"Oh—and Ethan?"

"Yes?"

"Be careful who you trust."

Ethan frowned slightly. "Why?"

Daniel returned to the window. "Because drift isn't always unintentional."

The EPAC Encounter

One week later—still in the first month of the work—EPAC was not an official department. But everyone knew the people who behaved like it existed:
- — They cut definitions to fit their needs
- — They optimized for local success
- — They pushed changes quietly
- — They defended their numbers fiercely
- — They disliked governance unless they controlled it

Ethan encountered one of them sooner than he expected.

Archer Collins—Senior Director of Commercial Performance—cornered him in the elevator.
"I heard you made quite an impression at the board meeting," Archer said casually.

"I just answered the question," Ethan replied.

"Mm." Archer smiled thinly. "That's the thing about answers. They can change the balance of a room."

Ethan said nothing.

Archer continued, "I saw the list of terms Priya is working on. Margin, backlog, order definitions… you're stepping into territory that affects how performance gets reported. That makes people… nervous."

"I'm not trying to make anyone nervous," Ethan said.

"Oh, you will," Archer replied lightly. "Not because you want to. Because clarity always does."

The elevator reached his floor.

Before stepping out, Archer turned.
"One piece of advice, Ethan—sometimes ambiguity is useful. Be careful cleaning it up too fast."

The doors closed. Ethan felt a chill he couldn't quite name.

Priya Tests Ethan Again

That afternoon, Priya asked Ethan to join her in the small windowless room used for governance workshops.

On the whiteboard were ten terms.

"Today," Priya said, "I want to see how you think under pressure."
She handed him a marker.
"Define the first three terms—margin, active customer, and backlog—in a way that Finance, Operations, and Sales can all sign."

Ethan froze. "That's… impossible."

Priya didn't blink. "Correct. Now do it anyway."

He walked to the board.

Margin.

He took a breath and wrote:

1. Margin (governing definition)
Net value created from revenue after deducting all associated costs, allocated and direct, as approved in cross-functional policy.

Priya tilted her head. "Why that phrasing?"

"Because 'governing' means this is the anchor definition," Ethan said. "Not the only one. Finance can still have a full P&L view. Ops can keep controllable margin. Sales can have customer-level margin. But this one belongs to the company, not any one function."

Priya's expression flickered—approval, not surprise.

"Continue."

He wrote:

2. Active Customer (governing definition)
An entity that has completed at least one billable transaction within the defined 12-month evaluation period.

He explained, "Twelve months is long enough for Manufacturing. Sales can still use six months internally if they want. But the governing definition must have consistency."

Priya nodded. "Backlog."

Ethan hesitated, then wrote:

3. Backlog (governing definition)
Confirmed orders not yet fulfilled, governed at SKU and delivery-window level.

"And why 'confirmed'?" she asked.

"Because half our backlog disputes begin with unconfirmed orders counted as backlog by Sales and rejected by Operations."

Priya smiled. "You see the fractures instinctively. That's why Daniel trusts you. But trust is not enough. You need discipline to match your instincts."

"I'm learning," Ethan said.

"Yes," Priya replied. "You are."

The First Internal Resistance

Two weeks into the work, the pushback began.
Not loud, not direct, but unmistakable:
- delayed responses
- missing data
- passive resistance
- "We're too busy this week"
- "That definition won't fit our process"
- "That's not how we've always done it"

The biggest resistance came from the places Ethan expected:
- Sales metrics
- Operations exceptions
- Commercial concessions
- Local finance adjustments

But the most subtle resistance came from Archer Collins.
Archer never opposed anything outright.
He simply kept asking questions designed to slow things down:
"Are we sure this is the right time?"
"What if we create new risks by forcing alignment?"
"What's the benefit of standardizing something that has worked locally for years?"
"Can we pilot definitions in one region first?"

He framed every challenge as caution.
But to Ethan, caution began to feel like sabotage in disguise.

Daniel's Quiet Guidance

One afternoon, after a particularly difficult alignment meeting, Ethan met Daniel in the hallway.

"Confusing day?" Daniel asked.

Ethan exhaled. "People are resisting. They're polite about it, but… they're resisting."

"Of course they are," Daniel said. "Alignment removes comfort. It removes interpretive wiggle room. It forces clarity."
He stopped.

"And clarity is accountability."

Ethan nodded slowly.

Daniel continued, "Your job isn't to win arguments. Your job is to expose drift. Priya's job is to govern drift. My job is to remove barriers when the time is right."

He leaned closer. "Not all resistance is harmful. Some is fear. Some is habit. Some is protection."

"And some is EPAC," Ethan said quietly.

Daniel's eyes narrowed slightly. "Be careful with that word. It's descriptive, not accusatory. Use it to understand, not to label."

"I understand."

"Do you?"

Ethan paused. "Yes. I think I do."

Daniel smiled faintly. "Good. Because the next few weeks will test that understanding."

The First Ten Terms Completed

By the end of the third month, the breakthrough came quietly—a timeline milestone.

After dozens of workshops, alignment sessions, and late-night debates, the first ten governing definitions were finalized and approved by:
- Finance
- Operations
- Sales
- Governance
- Strategy and
- IT.

Each definition had:
- a clear owner
- a version number
- a business rule
- data rules
- usage guidelines

- exception path
- sign-off log

It was, in every way that mattered, the first shared language DWL had ever had.

Priya gathered the small team in the governance room.
"This," she said plainly, "marks the beginning of stability."

Kara smiled for the first time in weeks.

Daniel nodded once, satisfied.

Ethan exhaled slowly, realizing how far they had come from that moment in the boardroom.

They had not solved the company.
But they had solved how to begin.

Reflection—The Awakening of Alignment
What we observed
Ethan stepped from observation into influence, recognizing how definitional drift shaped every contradiction in DWL's numbers. He saw resistance, ambiguity, and invisible forces protecting the status quo.

What it means
The company's challenges were not rooted in capability but in meaning. Without governing definitions, no metric could stabilize, and no transformation—technical or otherwise—could stand.

What will happen next
Move from defining terms to governing them.
The next phase will formalize The Language of Truth and prepare leadership to expand alignment from the first ten terms to the fifty ahead.

Aphorism
Transformation begins when someone finally sees the problem beneath the numbers.

Guiding Principle
Small definitional drift compounds into large organizational distortion.

Law—The Drift Amplification Law
Drift introduced upstream becomes disproportionately expensive downstream.

Equation
$$C_{rework} = k_2 \times D^2$$

Technical Explanation
Rework cost C_{rework} grows with the square of drift D.
A small increase in definitional drift produces a much larger increase in reconciliation time and cost.

Symbols
C_{rework}: Cost or effort spent on reconciliation and fixing.
D: Magnitude of definitional drift (how far current meaning deviates from the governed one).
k_2: Positive scaling constant translating drift into rework cost.

CHAPTER 3—The Language of Truth

A different kind of quiet

The first week after the board meeting felt strangely quiet. Not because there was less work, but because a decision had been made that no one yet knew how to carry out.

Daniel had drawn a clear line: no new analytics or AI initiatives until DWL could agree on what its most important words meant. That sentence had landed hard in the technology and analytics teams, softer in the commercial teams, and almost invisibly in the plants.

For Ethan, it changed everything. Until now, his work had been framed as "analysis" or "support." Now it carried a different label, unspoken but real: translate the company to itself.

The Language of Truth initiative began the way most important work does in large organizations—not with a public launch, but with a short, carefully worded meeting invite.

The subject line was simple:
"Language of Truth—Core Term Council (Initial Session)"

The invite list was small enough to fit around a single table. Daniel would chair the first session, but he had made it clear to Priya that this was her arena. She would own the structure; he would own the expectation. Ethan had a seat as well, not as a decision-maker, but as the person responsible for showing what misalignment looked like in the numbers.

No one outside that group yet called it a movement. Internally, it was framed as a—cleanup effort to tighten definitions, reduce reconciliation noise, make the next board reporting pack less painful. Only a few of them sensed that something more fundamental was beginning: a shift from everyone defending their own version of truth toward building one shared language they would all be measured by.

The following weeks carried a different kind of quiet—not tense, not defeated, but newly attentive.
The leadership team did not speak more than usual.
They listened more than usual.

For the first time in years, the organization was not rushing toward another initiative or scrambling to justify last quarter's projections. Instead, it had stepped into a rare pause—intentional, sober, and necessary.

By the second month-end, since the board confrontation, that pause was beginning to take shape as work.
Not technology.
Not tools.
Not dashboards.
Work made of words.

The First Convening

The invitation list was unusually short.
No extended teams.
No deputies.
No analysts except one.

Only:
— Daniel Shaw—CEO
— Elena Park—CFO
— Aaron Cole—COO
— Maya Chen—Head of Sales & Customer Operations
— Priya Nayar—Head of Governance & Risk
— Ethan Anderson—Senior Manager, Strategic Analytics

The group assembled in the smaller strategy room next to the main boardroom. That room had no automatic screensavers, no rotating dashboards, no wall of key indicators. It was intentionally plain—whiteboard, table, chairs, and the weight of expectation.

Daniel opened without preamble. "We agreed in the board meeting that we need a shared language. Today is the first step. We're not defining every term in the company. We're identifying which terms, if aligned, unlock everything else."

Priya added, "We're starting with the five terms that surface in every major decision: margin, order, active customer, backlog, and on-time."

"And three more," Elena said. "If we're serious about cross-functional alignment." She clicked her pen once, the way she did when she was switching from observation to strategy. "Recognized revenue, cost of goods sold, and adjusted forecast."

Aaron exhaled quietly. "That's eight."

Maya responded, "Eight that drive everything."

Ethan sat with a notebook open, not typing—writing. It helped him think more honestly.

Daniel looked around the room. "Before we jump in… does anyone believe this effort is optional?"

No one spoke. That clarity was new.

The Problem Beneath the Problem

Priya stood, moved to the whiteboard, and wrote in large letters:
"Order"
What does this word mean?
Underneath, she drew a line and wrote:
– To whom?
– When?
– For what decision?
She stepped back. "We assume shared meaning when we have shared spelling. That's our mistake."

Maya nodded. "Sales uses 'order' to mean customer commitment. Operations use it to mean a production instruction. Finance uses it to mean revenue recognition."

Aaron added, "And IT calls everything an 'order object.' Which means nothing outside their world."

Daniel looked at Ethan. "This is where you come in."
Ethan straightened. "During reconciliations over the last year, every time we traced a mismatch, it connected back to a definition gap. Same data, same systems, different meaning. We've been correcting symptoms, not the cause."

"Which cost us how much time?" Maya asked.

Ethan flipped back through his notes. "On average? About 200 hours per month across teams. Not including leadership time."

Elena raised a brow. "Two hundred?" "Conservatively."

Daniel let the number land. "That's not inefficiency. That's structural friction."

Priya added, "And it compounds. Every month that we operate with misaligned meanings, our data and reports drift further apart."

Aaron asked, "So what's the first term?"

Priya turned back to the board.
She wrote:
"Margin"
Eight letters. A thousand interpretations.

The First Definition Debate

Elena started. "Margin is full P&L—revenue minus total cost, including allocations. That's the standard for financial accuracy."

Aaron countered, "But allocations distort operational truth. I can't hold plant managers accountable for corporate overhead."

Maya added, "And customer-level concessions change the story entirely. What Finance calls 'discount leakage' is often the price of retaining strategic accounts."

For a moment, they were back to where they started—three truths in conflict. But this time, something subtle was different.
No one was defending their truth. Each was trying to understand the others.

Priya asked, "What if the problem isn't the number, but the expectation? Each function needs its own margin. But for cross-functional decisions, we need a governing definition."

Daniel nodded. "A single definition that we all accept as the company's shared truth. Not the only margin, but the one used when we speak to each other—and to the board."

Ethan leaned forward. "So, we define three versions: Financial Margin, Operational Margin, and Commercial Margin. Then one governing term: Enterprise Margin."

Elena paused. "Enterprise Margin?"

"Built from the reconciled components," Ethan said. "Not a compromise— an integrated view. And everyone agrees that Enterprise Margin is what we present when decisions span functions."

Aaron considered it. "Operationally workable. I'd still track my own version, but I won't argue with the integrated one."

Maya added, "As long as commercial adjustments are visible and not buried."

Daniel looked around the room. "Does anyone disagree?"

Silence.
Consensus—not by force, but by understanding.
The first definition took forty-two minutes.
And it changed the company's trajectory.

Ethan's Role Solidifies

As they moved to the next term—**Active Customer**—the conflicts sharpened.
Sales counted customers the moment they placed an order.
Operations counted them when manufacturing shipped.
Finance counted them when revenue was recognized.

Priya said quietly, "Three clocks. Three realities."

Ethan spoke up. "The simplest definition is: an entity for whom we delivered value in the last 12 months. But the operational trigger should match what we control."

Maya challenged him. "But Sales needs visibility earlier."

Aaron countered, "But Ops can't plan against customers who exist only in CRM."

Elena added, "And Finance can't report customers based on intent."

Daniel asked Ethan directly, "What's the reconciled truth?"

Ethan hesitated. "We need a two-tier definition. A Planning Customer—early-stage, intent-driven. And an Active Customer—value-delivered."

Priya nodded. "That gives each function what they need without breaking reality."

Daniel smiled faintly—approval without words.

Ethan felt something shift.
He wasn't just analyzing now.
He was shaping the foundation.

The awakening wasn't cinematic.
It was responsibility.

By the time the group met again,
the organization had entered what would later be called:
The third month of alignment.
No one said that out loud.
No one needed to.

It showed up in behavior.
— Emails became clearer.
— Cross-functional conversations became less defensive.
— Directors began asking, "What definition are we using?"
— Meetings shortened.
— Reconciliation hours dropped.
No dashboards had changed.
No software had been deployed.

But meaning was beginning to standardize.
This was the first visible return on the pause.

The Governance Rubicon

One afternoon, Priya brought up the unavoidable question.
"Once we define these eight terms… who owns keeping them aligned?"

Elena said, "Finance should own definitions tied to reporting."

Aaron said, "Operations should own anything related to throughput."

Maya said, "Commercial definitions shouldn't be overridden by Corporate."

Daniel intervened. "Ownership is not the same as alignment. Ownership answers 'who maintains the term.' Alignment answers 'who lives by it.'"

Priya added, "We need both."
After a pause, she drew a simple structure on the whiteboard:
— Governance owner
— Functional co-owners
— Decision rights
— Cross-functional enforcement

Aaron shook his head. "That sounds like bureaucracy."

Priya responded, "Only if it slows us down. Effective governance removes friction. Bad governance creates it. This approach removes friction."

Daniel said, "This is not about bureaucracy. This is about building the discipline we've avoided for years."

The group agreed—reluctantly, at first, then fully.

The Language of Truth would not just be a set of definitions.
It would be a **system**.
A living agreement across the company.

The Word That Almost Broke Them

The hardest definition wasn't "margin."
It wasn't "active customer."
It was: **On-Time**
It seemed simple.
It was not.

Operations measured "on-time" by production schedules.
Logistics measured it by shipment data.
Sales measured it by customer promise dates.
Finance tied it to service-level agreements.

Each definition was valid.
Each contradicted the others.

Daniel finally asked the question that mattered:
"Which definition best represents the promise we make to customers?"

Silence.

Then Maya said, "The customer's promise date. Full stop."

Aaron pushed back. "But if Sales sets unrealistic dates, Operations will always look like we failed."

Elena added, "And if we choose the customer promise date, we accept that commercial teams define operational reality."

Daniel stepped between them—not literally, but functionally.
"Let's think bigger. A promise is not just Sales. A promise is the coordination of commitments across all functions. A promise date that is impossible is not a customer commitment—it's a breakdown."

Priya added, "Then the governing definition of on-time must be:
Customer Promise Date validated through cross-functional feasibility."

Ethan said, "We can track the upstream and downstream components separately. But the governing metric is the validated promise."

Aaron nodded slowly. "Agreed. As long as feasibility checks are explicit."

Maya said, "Deal."

Elena added, "As long as reporting reflects the integrated version."

Consensus formed—again.
This time more easily.
The team was learning how to align meaning without defending territory.
That was the real transformation.

The First True Breakthrough

Three weeks later, the group reconvened with drafts of the eight definitions. Priya had converted their discussions into structured, audit-ready terms with:

- Definition
- Purpose
- Owner
- Co-owners
- Calculation logic
- Decision rights
- Exceptions
- Review cadence

When she displayed the consolidated table, something unexpected happened.

Maya exhaled with relief.
Aaron leaned back, calmer than usual.
Elena nodded, visibly satisfied.

Daniel said quietly, "This is the first time in my seven years here that I've seen all three of your worlds expressed as one."

The group wasn't celebrating.
They were settling in.
It felt like adulthood.

Ethan watched the others.
This was not a moment he would have noticed months ago.
But now it meant something.

This was alignment becoming culture.

Ethan's Internal Shift

In the next weeks, as the team refined the terms, Ethan found himself noticing patterns he had ignored before:
- How often teams solved the wrong problem
- How frequently meetings were spent repairing miscommunication
- How data disputes were emotional, not numerical
- How often leaders interpreted the same slide differently

He understood now why Daniel and Priya wanted him in the room.
Not just for analytics.
For perception.
He was starting to see the company the way they saw it—as a system built not on processes, but on understanding.
He did not feel like a hero.
He felt responsible.
And that was the point.

The Unspoken Shift in Leadership

By the end of Fourth Month, the leadership team had changed subtly.
Daniel became quieter but more precise.
Elena became more collaborative.
Aaron became less defensive.
Maya became more strategic.
Priya became more central.
Ethan became more involved.
None of this was declared.
There were no new titles.
No org charts changed.

But the organization had begun to move with shared intention.

That was the real metric.

A Tension Emerges

Just when the group seemed aligned, a subtle conflict surfaced.

Several Directors from middle management began pushing back.
Some were concerned about losing autonomy.
Some felt definitions imposed by Corporate were unrealistic.
Some feared visibility into their adjustments.
Some preferred ambiguity—it protected them.

This was the first sign of resistance.

Priya anticipated it.
Daniel accepted it.
Elena respected it.
Maya understood it.
Aaron prepared for it.
Ethan feared it.
Not because of the conflict itself, but because he now understood the stakes.

This was the moment when definitions became discipline.
When alignment became governance.
When understanding began to challenge habit.
This was the first hint of the deeper conflict to come.
Not between people and systems, but between truth and convenience.

The First Readout to the Extended Leadership Team

The room was full.
Directors.
Plant managers.
Regional controllers.
Commercial leads.
IT managers.

Daniel stood at the front.
"We're not here to debate the definitions," he began. "We're here to explain why we can no longer operate without them."

Elena presented the financial alignment.
Aaron presented the operational alignment.
Maya presented the customer alignment.
Priya presented the governance structure.

Then Ethan stepped forward to present something simple and devastating:
Before and After Reconciliation Hours.

He showed:
— 210 hours/month $\rightarrow$ 141 hours/month
— A 33% reduction in reconciliation work
— With only six definitions finalized
— Without a single technical change

The room went silent.

Not impressed—aware.

For the first time, middle management saw that definitions were not academic.

They were productive.

At the end of that meeting, Daniel said something that summarized the turning point:

"We used to think clarity came from better tools. Today we learned clarity comes from better agreement. The tools come later."

Priya added, "Alignment is not a deliverable. It is a practice."

Ethan said nothing. He just watched the shift in the room.

He had seen what misalignment did.

Now he was seeing what meaning could repair.

This was not the end of the work. It was the end of denial.

Reflection—What We Learned by Third and Fourth Months
What we observed
Definitions that seemed simple revealed deep cross-functional divergence.
The organization accepted, for the first time, that misalignment had been
draining trust and productivity.

What it means
Shared language is not documentation—it is operational infrastructure.
It reduces friction, unlocks collaboration, and prevents data drift.

What will happen next
Finalize the eight governing definitions.
Establish ownership, co-ownership, decision rights, and review cadence.
Prepare for the next phase: transforming definitions into discipline.

Aphorism
Truth begins with the courage to define it.

Guiding Principle
Memory without meaning is storage, not intelligence.

Law — The Law of Meaningful Memory
A system remembers clearly only when its inputs are semantically governed.

Equation
$I_{usable} = M_{raw} \times G_{meaning}$

Technical Explanation
Usable intelligence I_{usable} equals raw memory M_{raw} multiplied by meaning
governance $G_{meaning}$.
If meaning governance is weak or zero, additional data does not increase usable
intelligence.

Symbols
I_{usable}: Intelligence that can actually support decisions.
M_{raw}: Volume of raw stored data.
$G_{meaning}$: Strength of semantic governance on that data (definitions, ownership,
rules).

CHAPTER 4—Fifty Definitions That Began a Movement

This marks the shift from a leadership initiative into a company-wide movement.

The list began as a spreadsheet with more ambition than formatting. Fifty terms. Five columns. No agreement.

The initial "candidate terms" had come from everywhere—Finance, Operations, Sales, Supply Chain, HR, Compliance, and IT. Each function had been asked a single question:
"Which words, if misinterpreted, would cause us to make a bad decision?"

The answers were predictable and revealing. Finance sent a list full of "margin," "earnings," "accrual," and "provision."

Operations focused on "throughput," "downtime," and "on-time."

Sales cared about "active customer," "pipeline," "commit," and "churn."

IT added "source of record," "golden record," and "master data."

When Ethan merged the lists, duplicates did not collapse cleanly. The same words appeared multiple times with different descriptions attached. "Active customer" alone had four definitions. "Order" had six.

That was the moment he realized they were not building a glossary. They were exposing a map of disagreement.

Priya printed the list for the first workshop, even though everyone had tablets. She wanted people to feel the weight of the pages, not just scroll past them. On the first line of the agenda, she kept the framing blunt:

"Objective: Agree to one meaning per term that we are willing to govern and live by."

It sounded straightforward. No one in the room yet knew how much emotion a single word could carry once performance, incentives, and accountability were tied to it.

In the weeks following, something unusual happened inside DWL: people began speaking more slowly.

Not because they were uncertain, but because they were realizing how much their certainty depended on definitions they had never actually aligned.

Conversations that once rushed forward now paused at moments that used to pass unnoticed. It was subtle at first—a question here, a clarification there—but by third week, the change had taken on its own rhythm.

The Language of Truth initiative was no longer an idea. It had become work.

The first cross-functional session took place in DWL's mid-size strategy room—larger than a project space, smaller than a boardroom. It had a whiteboard that spanned an entire wall, round tables arranged in a semi-circle, and a ceiling speaker system that, on this particular morning, carried a mild hum from the HVAC. Nothing about the room signaled transformation. That made it perfect.

Priya and Daniel arrived first.

"This is where the friction will show," Priya said quietly.

Daniel nodded. "Better here than in front of the board again."

A few minutes later, the others filtered in: Elena from Finance with her binder of assumptions, Aaron from Operations with printouts of runtime metrics, Maya with her sales performance tracker. Analysts and managers joined them—people who handled reconciliations, pipeline data, plant logs, and customer records. Ethan took a seat near the center, laptop closed, ready to observe before contributing.

The room settled, though not comfortably. Everyone understood why they were there, but no one knew exactly how it would unfold.

Priya stood. "We're here to define the terms that decide how this company understands itself. We're not creating new metrics. We're exposing the inconsistencies in the ones we already use."

On the whiteboard behind her, three words were written:
- What
- Who
- How

She pointed to them as she spoke.
"What does this term mean?
Who owns it?
How is it used across functions?"

That was the structure. Nothing more. Nothing less.

A few heads nodded. Others looked cautious.

Daniel added, "Before we debate definitions, we acknowledge something up front: every function has built its own logic out of necessity. Finance didn't invent its view to challenge Operations. Sales didn't adjust terms to undermine Finance. Each team adapted definitions to meet their outcomes. That's normal. It becomes a problem only when individual adaptations are treated as universal truth."

Elena spoke next. "So, our job is to align where it matters and preserve differences where they're valid."

"Exactly," Priya said, relieved that Elena had set the tone.

They started with the term that had triggered the boardroom tension:

Margin.

Not to redefine it into a single number, but to understand why three versions existed in the first place.

Elena began. "Finance margin uses fully allocated costs. It reflects the economic reality of the company. It's not meant to be compared directly to operational performance."

Aaron responded calmly, not defensively. "Operational margin excludes allocated costs. It measures what I can control. If Finance's definition is the governing one, then my teams are accountable for items they don't have influence over."

Maya added, "And Sales margin factors in discounts and concessions that neither of you capture in real time. If a quarter requires them to save a renewal, our margin takes a hit that isn't reflected equally in your lenses."

Ethan listened as the patterns revealed themselves, more behavioral than technical. Each view was accurate—within its boundary. The problem wasn't the numbers. It was the assumption that a single term meant the same thing in every context.

Priya wrote three phrases under the word **margin**:
- Enterprise Margin (Finance)

-   Operational Margin (Operations)
-   Commercial Margin (Sales)

She circled them. "This is our first truth: multiple definitions exist because multiple purposes exist. So, the question isn't which one is 'correct,' but which one is 'governing' for each decision."

Silence, then agreement.

It was enough for day one.

Two weeks into the effort, the pace had accelerated.

The group met twice a week, sometimes three. The sessions grew more structured. Each meeting tackled a cluster of terms—revenue, order, backlog, on-time, inventory, forecast, cost-to-serve, active customer. They weren't trying to reach perfection; they were trying to uncover the assumptions that had been invisible for years.

Patterns emerged quickly.
Finance prioritized economic accuracy.
Operations prioritized controllability.
Sales prioritized customer impact.

These weren't disagreements. They were orientations. And once the group understood them, conversations shifted from debate to design.

By the end of the second week, they had identified fifteen terms that required alignment. By the end of the third, twenty-eight. Soon the list crossed forty, then approached fifty.

No one said the number out loud, but everyone felt the weight of it. These were not obscure technical labels. These were the terms that dictated decisions, funding, bonuses, performance reviews, and strategic choices.

Ethan saw another pattern—one the others hadn't named explicitly.
Departments used language to protect themselves.
Not intentionally. Not maliciously. But defensively.
When pressure built, definitions drifted.
When drift accumulated, truth blurred.
When truth blurred, trust decayed.
And once trust decayed, reconciliation became a permanent tax on time.

He captured the pattern in his notes:

Drift → Blurred truth → Lost trust → Structural friction → Reconciliation → Fatigue

What struck him most was how preventable it had all been.

But preventable didn't mean easy.

During one particularly difficult session, they tackled the term

Backlog.

Elena began, "Backlog is booked revenue not yet recognized."

Aaron countered, "Backlog is work committed but not yet delivered."

Maya offered, "Backlog is customer demand we've secured but haven't fulfilled."

Three definitions. Three realities. Three consequences.

"It's no wonder our numbers never reconcile," Aaron said, half under his breath.

Priya leaned forward. "Which version of backlog governs how we prioritize production?"

Aaron responded immediately. "Mine. The operational backlog reflects what we must physically process."

Priya turned to Elena. "Which version governs financial guidance?"

"Mine," Elena said. "Investors expect backlog to represent future revenue."

Priya shifted to Maya. "And which version governs customer communication?"

Maya smiled. "Definitely mine. Customers don't care about accounting treatments or production schedules. They care about commitments."

Priya stepped back. "Then our job is to map how these three definitions interact, not collapse them into one."

Ethan added, "We also need to document the interfaces. Backlog isn't one metric—it's three different signals that flow into each other."

Daniel nodded. "This is the work. Not solving friction. Understanding it."

By the end of the session, backlog had three accepted lenses—but also a shared rule: for any cross-functional decision, the governing definition would be explicitly chosen, not implicitly assumed.

It was a breakthrough—not because the term was resolved, but because the team learned how to resolve terms.

The turning point came during the fifth week.

The group was reviewing progress, mapping dependencies, and identifying terms whose definitions impacted multiple others.

They had just finished a challenging discussion about "cost-to-serve" when Priya noticed something.
"Look at the list," she said quietly. Everyone did.

"Fifty."

Not engineered. Not targeted. Not planned. Just the natural endpoint of all the terms the company had been using without alignment.

Daniel exhaled slowly. "We didn't set out to define fifty terms. We set out to define the ones that matter."

Elena added, "These aren't 'metrics.' These are decisions."

Aaron scanned the list. "And this map directly to our biggest reconciliation loops."

Maya nodded. "No wonder our stories never matched. We've been describing the business through different languages."

Priya wrote two words at the top of the board:

"Shared Meaning"

Ethan watched the room. No applause. No celebration. Just the recognition of a truth that had been waiting to surface.

Fifty definitions. Fifty agreements. Fifty commitments.

The movement had begun.

As alignment deepened, something shifted culturally.

Teams started asking better questions.

Analysts began challenging assumptions respectfully.

Directors started bringing definitional issues to meetings before they became escalations.

And leaders—people who had been in the company for years—admitted how much they had been operating on autopilot without realizing it.

It wasn't humility. It was clarity.

By the mid-second month, the initiative had created something unexpected: momentum.

People noticed that meetings were shorter. Cross-functional reviews felt less adversarial. Reconciliation cycles required fewer rounds. Conversations that used to take an hour took fifteen minutes.

Not because people worked faster.

Because they finally meant the same thing when they used the same words.

But progress did not come without resistance.

One morning, during a review with the finance controllers, the resistance surfaced plainly.

A controller pointed to a definition the group had recently aligned. "If we adopt this, we'll need to redo three years of comparatives."

Elena responded gently but firmly. "We don't need to redo history. We need to stop rewriting it."

A pause. A long one. "It's not that we don't agree," the controller said quietly. "It's that if we publish this version, past performance will look different."

Ethan understood the tension. "Definitions are not retroactive punishments," he said. "They're forward-looking agreements. The goal isn't to make the past cleaner—it's to make the future less confusing."

The room softened.

The resistance was not ideological. It was human. The fear that alignment would expose past inconsistencies was natural. But the alternative—continuing until every inconsistency became unmanageable—was worse.

Priya framed it simply. "This is not about correction. It is about standardization."

From that point on, resistance did not disappear. But it stopped slowing the group down.

They weren't trying to repair the past.

They were protecting the future.

The moment the initiative became real for the entire company happened quietly, during a monthly leadership call.

Each function presented updates. The tone was routine until Daniel asked a question none of them had anticipated. "For your metrics today," he said, "which governing definitions are you using?"

There was a pause, then something remarkable.

Each leader referenced the same definition document.
Not because they had been told to.
Because they now trusted it.

Ethan watched the shift with a mix of pride and caution. This was the beginning, not the end.
Alignment was not a project.
It was a discipline.
A way of operating.
The fifty definitions were not an artifact.
They were a commitment.

By the end of the third month, the company had changed in ways that were visible but not dramatic.
Meetings were calmer.
Performance reviews were clearer.
Leadership was aligned in language, if not yet in behavior.

And the organization—slowly, steadily—was becoming ready for what would come next.

With definitions stabilized, data began stabilizing.
With data stabilizing, systems behaved predictably.
With systems predictable, the conversation about advanced analytics and AI—which had been paused intentionally—could finally begin to re-enter the narrative later in the journey.
But not yet.

The foundation was still settling.

High maturity requires slow beginnings.

Reflection—When Meaning Becomes Momentum
What we observed
The company surfaced fifty unaligned definitions that governed how decisions were made. Through collaborative sessions, each term became understood, owned, and mapped to its purpose. Friction began decreasing as shared meaning increased.

What it means
Truth is not the result of data; it is the result of agreement.
Alignment is not a one-time exercise but an operating discipline.
When definitions stabilize, trust stabilizes.
When trust stabilizes, performance accelerates.

What will happen next
Finalize the governance model for these definitions.
Build the structures that will keep alignment intact as the company grows.
Only after this foundation is secure will DWL explore the responsible introduction of advanced analytics and AI.

Aphorism
Clarity is not created by tools. It is created by people willing to agree on meaning.

Guiding Principle
A definition is a control system, not a vocabulary term.

Law — The Term Integrity Law
A term is valid only if ownership, lineage, and usage remain consistent.

Equation
$T_{integrity} = O \times L \times U$

Technical Explanation
Term integrity $T_{integrity}$ is the product of ownership clarity O, lineage clarity L, and consistent usage U.
If any of these is missing or near zero, the term cannot reliably support reporting or AI.

Symbols
$T_{integrity}$: Overall integrity of a business term (e.g., "margin", "order").
O: Strength of explicit ownership (who is accountable).
L: Quality of lineage understanding (where the term comes from and how it's transformed).
U: Consistency of usage across functions and systems.

PART II—THE RECKONING

Where DWL confronts its invisible debt—semantic, operational, and cultural. Alignment collides with habit. Governance begins to take shape.

CHAPTER 5—Building the Governance Backbone

The impact of the first fifty definitions was immediate—subtle, but unmistakable.

The first fifty definitions had changed the temperature of the company.
Not dramatically.
Not loudly.
But visibly.
Teams no longer argued over the meaning of "margin" in every meeting.

Sales and Finance had stopped using competing versions of "active customer."

Operations had finally aligned its plant-level metrics with corporate definitions that had once seemed irrelevant.

The friction wasn't gone, but the ricochet effect—the instinct to reinterpret definitions on the fly—had slowed.

That was enough for Daniel Shaw to move to the next stage.

He had always known that definitions alone were not the destination.
Words were the first layer.
Behavior was the second.
Governance was the third.
Without governance, definitions would slip back into entropy.
Without governance, the fifty terms would remain policy statements rather than operating principles.
And without governance, the company could never introduce advanced analytics—let alone AI—without repeating the mistakes of the past.

This was the moment the Governance Backbone needed to be built.
Not as bureaucracy, but as discipline.
Not as oversight, but as alignment.
Not as control, but as clarity.

The First Gathering of the Backbone

The meeting took place in DWL's mid-floor conference room—the one that felt large when only two people were inside it, and strangely small when more than ten were.

Daniel wanted this meeting to feel intentional.
Not rushed.

Not improvised.

Just enough formality to signal seriousness; just enough informality to avoid defensive posturing.

At the table sat:
- **Priya Nayar**, Governance & Risk—newly positioned as the architect of meaning across functions
- **Elena Park**, CFO—the anchor of financial truth
- **Aaron Cole**, COO—responsible for operational grounding
- **Maya Chen**, Sales & Customer Operations—voice of the customer and commercial integrity
- **Ethan Anderson**, Strategic Analytics—pattern-recognition, emerging leadership

And three new faces invited for global scope:
- **Rosa Martins**, EMEA[6] Finance Director
- **Kenji Watanabe**, APAC[7] Operations & Quality Lead
- **Luis Herrera**, Americas Commercial Controller

This was the first time global leaders were brought into the inner structure of the Language of Truth initiative. Not for symbolism, but because definitions meant nothing if they couldn't survive scale.

Daniel opened the meeting.

"Thank you all for being here. The work we've done so far—those fifty definitions—has helped stabilize the language across our core functions. But to move from alignment to sustainability, we need structure. Today isn't about reviewing terms. It's about deciding how we govern meaning going forward."

He paused, letting the phrase "govern meaning" settle.

It was not a typical corporate phrase. But it was accurate.

"This is the beginning of what we'll call the Governance Backbone," Daniel continued. "Definition alignment is the first step. The next step is discipline—how we maintain alignment as the business changes."

He nodded to Priya. "You'll lead us through it."

The Framework That Would Hold the Company Together

Priya stood and moved toward the front monitor.

"This isn't a committee," she began. "It's an operating system."

---

[6] **EMEA**—Stands for Europe, Middle East, and Africa It's a business and geographic designation used by global companies to group these regions together for organizational, marketing, and reporting purposes

[7] **APAC**—Stands for Asia-Pacific It's a regional designation used in business, economics, and geopolitics to group countries in East Asia, South Asia, Southeast Asia, and Oceania under one umbrella

Ethan glanced at her—not because the analogy was dramatic, but because it was precise. Priya was not prone to metaphors; she chose her language deliberately.

She clicked into a slide labeled:

"Governance Framework v1.0—Roles and Responsibilities"

"Every definition," Priya said, "needs three roles to survive beyond the moment it is agreed."

1. Owner
   – Responsible for the meaning and the authoritative semantic direction of the term.
   – Makes final semantic decisions.
   – Usually, a VP or senior director.

2. Steward
   – Responsible for maintaining the term in practice.
   – Ensures the definition is applied consistently.
   – Usually, a functional expert.

3. Custodian
   – Responsible for the data representation.
   – Ensures system fields and pipelines accurately reflect the definition.
   – Usually, analytics, data engineering, or IT.

Priya continued:
"When we began reviewing definitions, we saw the same pattern everywhere: no one knew who owned what.
People assumed.
People improvised.
And because the owners weren't explicit, definitions drifted silently."

She looked around the room.
"This structure is how we stop the drift."

Global Leaders React

Rosa spoke first. "Will each region follow the same roles and ownership? Or will ownership only be at corporate?"

Priya nodded. "Ownership is global. Stewardship can be regional."

Kenji added, "And what if a term has different operational relevance in APAC?"

"It will," Priya said. "But relevance is not the same as meaning. The meaning must be the same globally. Application can differ."

Luis leaned forward. "We've needed this for years. Controllers feel the impact of drift more than anyone. Every quarter close becomes a negotiation. This will reduce a lot of that noise."

Daniel noted the reactions silently. These were leaders who dealt with consequences—not slides.

This was the first sign the Governance Backbone would hold.

Ethan's Quiet Role Shift

When Priya finished explaining the structure, Daniel asked Ethan to speak. "From the analytics side," Daniel said, "what do you see?"

Ethan stood, feeling again the slight tension of being the least senior person at the table. But he also sensed this room trusted his analysis.

"We've mapped the fifty terms across fifteen major reports," Ethan said. "The biggest gaps aren't in the definitions—they're in the data structures behind them."

He clicked into a dashboard.

"Look here," he said. "Three regions use the same term 'active order,' but they map it to different fields. APAC uses a field that counts any order with activity in the last 60 days. EMEA uses 90 days. Americas uses 30 days."

Maya frowned. "We agreed it was 30."

"We did," Ethan said, "but the systems weren't updated. Which means no matter how aligned our definitions are, the reports will remain unaligned."

Aaron leaned back. "So where do we start?"

Ethan answered. "With the Governance Backbone. We assign owners, stewards, custodians. Then we take the top ten definitions that show the highest global variance and bring their data representations into sync."

"Top ten?" Rosa asked.

"We'll eventually do all fifty," Ethan said. "But the top ten cause 80% of the noise."

Daniel nodded approvingly. Prioritization was the difference between motion and progress.

The Governance Backbone Becomes Real

For the next two hours, the group worked through the fifty definitions—assigning owners, stewards, and custodians for each.

Some decisions were easy:
Margin → Owner: CFO
On-Time → Owner: COO
Active Customer → Owner: Head of Sales & Customer Operations

Some were harder:
Order Start → split between Operations and Sales
Asset Utilization and Availability → global vs regional complexity
Forecast Accuracy → analytics implications

And some were political:
Revenue Recognition Date
Exception Classification
Product Activation

It became clear why governance had been avoided for so long.
It wasn't because people resisted clarity.
It was because clarity requires courage.

The First Signs of Resistance

Halfway through the session, Maya noticed a pattern.

Every time stewards were assigned for definitions that touched EPAC processes, representatives from Operations seemed uneasy.

At one point, while discussing "Exception Category," Aaron shifted uncomfortably.

"Operations own this term," Priya said.

Aaron hesitated—just a fraction of a second, but long enough for everyone to see.

"What's the concern?" Daniel asked.

Aaron exhaled. "Exception metrics get weaponized. If we own the term, we'll be held responsible for every exception, including those caused upstream."

Luis spoke gently. "But without ownership, no one is responsible. And that's how the drift keeps happening."

Aaron considered the point, then nodded reluctantly. "Fine. Operations will own it. But we need stewards from Finance and Sales too."

"Agreed," Priya said.

It was a small moment, but Ethan saw the psychology behind it.

Governance was not about rules. It was about confronting the places where power and accountability meet.

Building the First Governance Cycle

Once ownership was established, Priya revealed the next component. "This Governance Backbone isn't a one-time exercise. It's a cycle."

She displayed a diagram labeled:

Governance Cycle—Quarterly Rhythm:
The cycle included:
- Definition Review (only if triggered by change in process, product, or policy)
- Data Alignment Review (custodians confirm system representations match definitions)
- Variance Report (analytics team reports deviations to owners)
- Decision Log Update (every semantic change must be recorded)
- Leadership Review (Daniel + CFO + COO + Sales + Governance)

"This," Priya said, "is how we keep Language of Truth from becoming a one-year project. It becomes part of how we run the company."

Daniel added, "And every quarter, the board will hear one version of each number—not three."

By Month Five of the transformation—the entire leadership team sensed the shift:
Fewer debates over meaning
Fewer reconciliations
Fewer escalations
More confidence in reports

More predictable variance patterns

Teams still disagreed.
But they now disagreed from the same starting point.

That alone saved hours each week.
It also freed emotional and cognitive space for the next phase of the transformation.

Ethan's Second Awakening (Quiet but Important)

After the meeting ended, Ethan stayed behind to review the mapping between definitions and data structures.

Kara joined him. "You looked more comfortable today," she said.

"Not comfortable," Ethan replied. "Just… clearer."

"About what?"

"That alignment is a leadership problem disguised as a data problem."

Kara smiled. "Welcome to governance."

He paused, then added quietly, "I think Daniel and Priya want me more involved."

"They do," Kara said. "Just don't underestimate the pushback."

"From who?"

"You'll see," she said. "EPAC never disappears easily. It adapts."

It was not a warning. It was a fact.

And Ethan understood that the Governance Backbone he was helping build would need reinforcement—not just process reinforcement, but cultural reinforcement—across the transformational timeline that would unfold over the next eighteen months.

The Governance Backbone Is Approved

At the next executive leadership session, Daniel summarized the work.
"The definitions gave us clarity," he said. "The Governance Backbone will give us endurance. Over the next few quarters, this will prevent drift, reduce

reconciliations, and stabilize the foundation we need before we introduce advanced analytics or AI."

No one argued.

The shift was quiet but unmistakable.

For the first time, DWL was not only documenting meaning—it was governing it.

And with that, the Language of Truth initiative moved from **movement** to **infrastructure**.

Reflection—The System Behind the Words
What we observed
The first fifty definitions created alignment, but not sustainability.
Ownership gaps, stewardship confusion, and data inconsistencies made it clear that meaning required structure—not intention alone.

What it means
Governance is not bureaucracy.
It is the operating system that prevents drift.
Without explicit roles and a quarterly rhythm, the organization would slide back into semantic fragmentation.

What will happen next
Launch the Governance Backbone formally.
Begin top-ten alignment of high-variance definitions.
Prepare for eventual re-introduction of analytics and early AI prototypes once the foundation is stable.

Aphorism
Definitions remove confusion. Governance removes recurrence.

Guiding Principle
A system can only correct what it can see.

Law — The Visibility Law
Reconciliation accuracy rises with visibility into the full transformation path.

Equation
$$R_{accuracy} = f(V_{lineage})$$

Technical Explanation
Reconciliation accuracy $R_{accuracy}$ increases as lineage visibility $V_{lineage}$ increases. Here $f(\cdot)$ is an increasing function: more visibility leads to better, faster reconciliation.

Symbols
$R_{accuracy}$: Accuracy and reliability of reconciliation outcomes.
$V_{lineage}$: Degree of end-to-end transparency of data flows and transformations.
$f(\cdot)$: A monotonic increasing function (higher input → higher output).

CHAPTER 6—The Pilot Illusion

When the company attempts to run before it can walk.

The first AI pilot at DWL didn't begin with a strategy. It began with impatience.

By Month Five, The Language of Truth initiative had aligned eight core terms—enough to stabilize reporting, reduce reconciliation cycles, and quiet the weekly debates between Finance and Operations. But the executive team, like most executive teams, was already looking beyond definitions.

They wanted acceleration—visible acceleration.

No one said it out loud, but the sentiment was clear:
"We've done enough alignment. Now let's do something that looks like AI."

The pressure didn't come from internal teams alone. A board member had forwarded an article about a competitor using AI for predictive operations. Another had asked Daniel privately whether DWL's AI roadmap was "keeping pace." And a large software vendor had been lobbying hard, suggesting a "quick win pilot" to build momentum.

Optimism, pressure, and the desire to show progress converged into a single decision:
Run one small AI pilot.
Something safe.
Something that proves potential.

Priya was careful about the framing.
Ethan was cautious.

Daniel approved it anyway—knowing the organization needed to see for itself what AI could and could not fix.

The choice of pilot was predictable: **predictive maintenance**.

Safe. Familiar. Bounded.
No cross-functional politics.
No direct revenue implications.
Clear data sources.
A problem everyone understood.

On paper, it should have worked.

The Pilot Begins Too Early

The vendor arrived with confidence.
A small team of consultants walked into DWL's innovation room carrying diagrams, templates, and a polished seven-week pilot timeline.

They brought their language, their model assumptions, their mappings, and their enthusiasm.

Some of the assumptions were valid.
Some were foreign.
Some were incompatible with DWL's newly forming definitions—but the vendor could not have known that.

The pilot kickoff lasted three hours.
The message landed clearly: *"We'll clean the data as we go."*

To the vendor team, that meant efficiency.

To Priya and Ethan, it meant danger.

Ethan raised his first flag gently.
"We're still solidifying definitions for downtime categories and asset hierarchies," he said. "The model might pick up inconsistencies as if they're meaningful patterns."

The vendor smiled politely. "We'll engineer around it."

Priya added, "Engineering around drift is how drift becomes permanent."

But the pilot had already begun. The company wanted action.

What the Model Saw—and What It Misinterpreted

Two weeks in, the model produced its first prediction.

A specific machine in Plant 3 was flagged as "high likelihood of failure in the next 48 hours."

Operations reacted immediately.
Aaron dispatched a maintenance lead.
The technician inspected the machine, ran diagnostics, and reported back:

"Nothing wrong. Machine's fine."

Operations concluded the model was overreacting.

The vendor disagreed.
Finance asked whether the flagged downtime would impact targets.

Sales ignored the alert altogether.

That divergence of response was the first hint that the pilot was not just a technical experiment—it was a mirror.

The real issue surfaced a week later.

The model flagged a different asset.
Then another.
Then an asset that had already been serviced.

Ethan pulled the model output apart line by line.
He found the patterns.
They weren't predictive.
They were definitional.

The model was "learning" from:
—   inconsistent downtime classifications
—   silent reclassifications across shifts
—   asset IDs that changed mid-quarter
—   maintenance logs written differently by different teams
—   flags that were sometimes callbacks, sometimes not
—   tickets that were closed prematurely to improve metrics
It wasn't that the model was wrong.

It was absorbing precisely what the organization had been feeding it.

Priya summarized it succinctly:
"The model is predicting the behavior of our data—not the behavior of our machines."

Daniel did not disagree.
But he let the pilot run longer.

Some lessons require pain to stick.

The Pilot Fails—But Reveals Something More Important

Week five ended with an internal review session.
The vendor presented the pilot results. The deck was visually perfect—charts,

heat maps, accuracies, confidence intervals, trendlines.

The story behind the slides was not.

Ethan began the discussion.
"Before we discuss accuracy," he said, "we need to look at the inputs. The model is interpreting inconsistent classifications as if they reflect machine behavior. It's capturing variation in human reporting as if it's mechanical degradation."
He switched to a simple graph.
Eight lines.
Eight definitions.
All labeled "downtime."
None matched.

A long silence followed.

Maya was the first to speak. "So, we built a model on top of eight different truths."

Aaron folded his arms. "And we're surprised it failed?"

The vendor attempted to defend their approach.
But Priya intervened gently. "This is not the vendor's fault," she said. "This is our system's reflection. The pilot exposed misalignment that maintenance teams have been compensating for manually. The model just made it visible."

Daniel nodded slowly. "This is not an AI issue. This is a meaning issue."

The vendor was thanked. The pilot was closed.

The team expected frustration from the board.

But Daniel framed the failure differently:
"The pilot succeeded. It proved we are not ready. It revealed the cost of misalignment better than any audit."

And privately, Ethan realized something else:
The company had not failed an AI pilot.

The pilot had succeeded—by revealing the truth.

The Psychological Impact—and Ethan's Turning Point

The pilot's collapse changed something inside Ethan.

He had seen misalignment before—definitions, reports, reconciliations. But this was different.

This time, the misalignment wasn't just causing confusion.
It was generating false confidence.
False urgency.
False signals.

He understood now what Daniel and Priya had sensed much earlier:

"AI amplifies meaning—whatever meaning exists, good or bad."

If DWL delivered misaligned inputs, the model would amplify the misalignment.
If DWL delivered definition drift, the model would treat drift as truth.
If DWL delivered inconsistencies, the model would treat inconsistencies as patterns.

It was the first time Ethan felt the weight of future responsibility—not as an analyst, but as someone who now saw the structure behind the noise.

And for the first time, he understood the deeper significance of the question Priya had asked him months earlier:

"Do you want to help us fix data, or help us fix meaning?"

He understood now.
There was no difference.

The Company Learns Its First AI Lesson

The Leadership Team met again two days later.

Daniel asked a single question: "Should we run another pilot?"

Priya: "Not yet."

Elena: "Not until downtime, asset hierarchy, and cost classifications are aligned."

Aaron: "And until the terms in the maintenance logs match the terms in Finance."

Maya: "And until we stop calling the same customer active in one system and inactive in another."

Daniel looked at Ethan.

Ethan spoke carefully.
"I think we should continue The Language of Truth until the core terms are stable enough that a model won't misinterpret human inconsistency as mechanical behavior."

Daniel nodded.
"This is the pilot's real value," he said.
"It showed us that alignment is not optional. It is prerequisite."

He paused. "And it showed us who we need to become."

The Alignment Reset

Daniel closed the meeting with a quiet final sentence:
"Before intelligence can help us, we must earn the right to use it."

The room did not respond with applause or frustration.
Just understanding.

The pilot had not damaged morale.
It had clarified responsibility.

The next phase of DWL's journey would not be defined by optimism or tools. It would be defined by discipline.

And as Ethan walked out of the room, he felt—more than ever—that the company had reached a threshold.

Not of technology.
Of maturity.

Tomorrow, the work of reconstruction would begin.

The next 12 months would be about correcting upstream signals, aligning behaviors, and preparing the ground for intelligence that could finally be trusted.

This was the last day DWL would mistake action for progress.

Reflection—What the Pilot Revealed
What We Observed
The first AI pilot failed not because the model lacked sophistication, but because the data reflected eight different operational realities. The pilot made misalignment visible.

What It Means
AI does not fix data.
AI does not fix definitions.
AI amplifies whatever meaning exists.
Misalignment becomes prediction error.
Inconsistency becomes learned behavior.

What will happen next
Freeze all AI experimentation.
Stabilize downtime definitions, asset hierarchies, customer states, and exceptions.
Ensure governing terms flow through logs, systems, and reports.
Restore consistency before reopening the AI roadmap.

Aphorism
A model is not a mirror of the future. It is a mirror of the present—polished with mathematics.

Guiding Principle
Preventing errors upstream is cheaper than correcting them downstream.

Law — The Preventive Control Law
Prospective controls reduce downstream noise exponentially.

Equation
$$N_{downstream} = N_{upstream} \times e^{-k_3 P}$$

Technical Explanation
Downstream noise $N_{downstream}$ falls exponentially as prospective control strength P increases. Even modest upstream controls can dramatically reduce exceptions, rework, and variance later.

Symbols
$N_{upstream}$: Baseline noise or errors introduced at the source.
$N_{downstream}$: Residual noise observed in downstream systems and reports.
P: Strength and coverage of prospective controls (validation at entry, guardrails, rules).
k_3: Positive constant controlling how quickly noise decays as controls strengthen.
e: Base of the natural logarithm (≈ 2.718), used for exponential decay.

PART III—THE MECHANISM

DWL shifts from defining truth to engineering it.
Real-time alignment, behavioral visibility, and the first dynamic corrections emerge.

CHAPTER 7—The Memory Lake

Where Truth Meets Structure

By Month Six, the Memory Lake did not begin as a breakthrough. It began as a quiet, necessary decision.

DWL needed a single place where aligned definitions, corrected values, reconciled histories, and clean data could coexist—stably, versioned, and visible.

A place that reflected not data exhaust, but data intention—the choices people made, not just the values they entered.

A place that showed the company not only what it had, but what its data meant.

They called it the Memory Lake because it preserved memory—not just of events, but of meaning.

What the Memory Lake Really Is

The Memory Lake was not a data lake.
Not a marketing concept.
Not a magic AI engine.

In plain English, it was built from:
1. Aligned Definitions
–   Every term passed through the Language of Truth framework.
–   Meaning was governed, not inferred.
–   A master term registry that controlled usage.

2. Corrected Histories
–   Past inconsistencies were recorded and versioned, not overwritten.
–   Silent changes became visible.

3. Timeline Sequencing
–   Each data object carried a timestamped lineage[8]: when it was created, updated, adjusted, or contested.

4. Behavioral Metadata[9]
–   Metadata describing how humans interacted with data—not just the data itself.

---

[8] **Lineage**—shows where data comes from, how it moves, and how it changes across systems. Technically, the record of data origins, transformations, and usage that ensures traceability and compliance across processes. for example, In Azure Purview (data governance service) and AWS Glue Data Catalog (index of metadata), lineage tracks customer order data from source systems through backlog reports, ensuring asset decisions were based on consistent information

[9] **Behavioral Metadata**—Metadata describing *how humans interacted with data*—not just the data itself. for example, How often a field changed, how fast, by whom, and whether rules were followed.

5. Unified Storage Layer
- Built in Azure (storage + metadata)
- Harmonized through pipelines
- Governed through traceable, rule-based transformations

This made the Memory Lake the first place in DWL where:
- the same term meant the same thing everywhere
- timelines were intact
- upstream adjustments were visible
- behavioral inconsistencies surfaced
- meaning was structurally preserved
It was a truth system, not a storage system.

How the Lake Begins to Identify Behavioral Patterns

The first insight came from a simple request.

Priya asked Ethan to run a historical comparison between two quarters, using the newly aligned definitions and corrected histories.

Ethan noticed something odd.
The numbers aligned—beautifully.
But the **rate of correction** did not.

Some teams corrected values at the beginning of the month.
Some corrected at the end.
Some corrected repeatedly for the same terms.

The Memory Lake didn't just store the corrected value—it stored the timeline of how that correction happened.

This revealed a new layer: correction patterns—how people changed data over time.

How the system derived behavioral patterns

1. Feature Vector Creation
   For each data object, the system generated a small, structured representation (a feature vector) containing:
   - number of changes
   - time between changes
   - who made each change
   - which rule triggered the change
   - whether a rule violation occurred

    – value distribution over time

2. Timeline Encoding
   Timeline encoding preserved the order of events:

   $$created \rightarrow updated \rightarrow reviewed \rightarrow adjusted \rightarrow finalized$$

3. Sequence Comparison
   The system compared sequences across teams:
   – Are Sales teams adjusting key fields late in the month?
   – Are Operations team revising the same fields repeatedly?
   – Is Finance applying corrections after period close?

4. Cluster Detection
   Using simple unsupervised clustering (no machine learning or 'AI magic'):
   – Teams who showed similar correction behavior clustered together
   – Outliers were visible
   – Patterns emerged across geography, function, or role

Result
The Memory Lake did not infer intent—it revealed behavior.

Not psychology.
Not motive.
Just patterned action.

How the Lake Reconstructed Meaning Over Time

Meaning is not only what is recorded.
It is when it is recorded, how often it changes, and what conditions triggered the change.
The Memory Lake preserved:
– every correction
– every adjustment
– every version
– every timestamp
– every ownership shift
– every transformation rule applied
– every pipeline change
– every business rule violation

The insight was simple:
"Inconsistency in meaning is visible through inconsistency in behavior."

For example:
- If "Active Customer" changed definitions mid-quarter,
  the timeline showed a sudden shift in classification patterns.
- If "On-Time Delivery" drifted between teams,
  the correction timeline showed bursts of manual adjustments.
- If downtime was misclassified,
  repeated category flips became visible across shifts.

Ethan realized DWL had never seen its own operational movement before.

The Memory Lake showed motion.
And motion revealed meaning.

When the Lake Revealed What No One Expected

It began with a single visualization.

Ethan showed Priya a timeline graph of adjustments across all regions.
North America showed smooth correction curves.
Europe showed end-loaded adjustments.
APAC showed repeated revisions to the same objects.

Priya stared at the graph.
"These aren't performance patterns—they're behavioral patterns."

The next surprise came from Finance.
Elena discovered her team was adjusting cost allocations more often than she realized—small changes but consistently timed right before internal reviews.

Operations discovered something else.
In Plant 4, the same downtime field was flipped six times in two hours—across shifts.

The Memory Lake didn't accuse—it revealed.
And leaders felt something unsettling:
For the first time, the company could see the hidden human layer behind the numbers.
Not who made errors.
But how meaning shifted, how definitions were applied inconsistently, how rules were bent by habit, workload, or interpretation.

Aaron summarized it bluntly:
"We're not dealing with data issues—we're dealing with behavioral residue."

When the Lake Began Anticipating Meaning Before People Did

DWL discovered something important:
This is not predictive AI—it is timeline-based pattern projection.

How it works is simple and accurate. Using the behavioral metadata + timeline sequences + frequency of adjustments, the system could:
  – detect when a team was likely to make late-cycle adjustments
  – identify which terms were at high risk of definitional drift
  – highlight repetitive transformation requests that signal process instability
  – forecast which data objects might require downstream correction
  – flag which processes were slipping toward manual workarounds

This was not "prediction" in the colloquial AI sense.

It was forecasting based on repeatable human patterns.

The key insight was discovered:
Patterns repeat.
Meaning drifts.
Teams behave in cycles—often in ways they do not notice.

And for the first time, the Memory Lake showed those cycles.

Priya called it 'behavioral visibility'.

Daniel called it 'the first real mirror'.

Ethan called it "The beginning."

Because once the company could see how its behavior shaped its data—it could begin shaping its behavior.

By the end of fifth month, the Memory Lake had stabilized enough for leadership to trust its internal logic—even if they were not yet ready to act on everything it revealed.

Patterns were no longer frozen in reports—they were moving, updating, and aligning in near-real-time.

It was the first hint that truth at DWL was not a destination—but a flow.

Ethan sensed that the next stage would not be about defining terms, but about managing the consequences of seeing them change.

The static world of definitions was ending.

The dynamic world of corrections was about to begin.

And that was the doorway to truth in motion.

Reflection—Seeing the System as It Truly Operates
What We Observed
The Memory Lake revealed that DWL did not simply have inconsistent data—it had inconsistent interpretations of time, responsibility, and operational reality. Timeline shifts, definitional ambiguities, undocumented adjustments, and subtle behavioral patterns surfaced clearly once meaning and data finally shared the same structure.

For the first time, leaders could see not just what changed, but how and when it changed.

What It Means
Misalignment was not a technical issue—it was an operational one, reinforced by habits, incentives, and silence.

The lake showed that definitions were not drifting randomly—they were drifting predictably, tied to pressure cycles, quarterly targets, and organizational shortcuts that had accumulated quietly over years.

Truth was not fragile—it was reactive.

What will happen next
A shift from static reconciliation to dynamic correction—a system that adjusts as the business moves.

This requires:
– a single place to log cross-functional discrepancies
– a mechanism to trace who altered meaning and why
– a way to align corrections across systems without reintroducing drift
– progressive stabilization of upstream systems so changes are not absorbed in pipelines

Introduce the Reconciliation Ledger—where truth stops being passive and begins to move with the business.

Aphorism
Clarity isn't found in the data. It's found in the distance between what changed and when we finally noticed.

Guiding Principle
A unified memory reveals patterns that fragmented systems hide.

Law — The Pattern Emergence Law
Coherent memory increases the discoverability of latent organizational patterns.

Equation
$P_{emergence} = k_4 \times C_{coherence}$

Technical Explanation

Pattern emergence $P_{emergence}$ scales with memory coherence $C_{coherence}$. The more unified and consistent the memory, the easier it is to see real behavioral patterns.

Symbols

$P_{emergence}$: Strength and clarity of patterns detected in data.

$C_{coherence}$: Degree to which data across systems is harmonized and consistent.

k_4: Positive constant translating coherence into pattern visibility.

CHAPTER 8—Truth in Motion

The Moment Flow Replaced Static Control

By Month Six, DWL was no longer a company that only looked at the past. It was starting—quietly and a bit awkwardly—to feel the present.

The Language of Truth work had stabilized the first fifty governing definitions.

The Memory Lake had begun absorbing clean, reconciled data.

Teams were learning to trust what they saw.
But the next challenge wasn't clarity.
It was motion.
Static definitions could stabilize yesterday.

What DWL needed now was a way to handle today, as it unfolded in real time.

This shift came to light during the first cross-functional performance review that used the newly aligned metrics. The dashboards showed synchronized numbers for the first time in years. Margin, backlog, asset uptime, and customer renewals all presented the same values across Finance, Operations, and Sales.

But Daniel noticed something else.
The alignment was clean at the start of the month—then began drifting by Week 3.

At first the deviations were small: a few basis points in margin, a slight timing difference in backlog, a new interpretation of "active order" in one regional team.

By the end of the Month-Six review, the deviations were visible enough that Elena said what the entire room was thinking.
"We aligned the definitions," she said quietly, "but we haven't aligned the movements."
This was the moment Priya had been waiting for—not because she wanted friction, but because without friction, there was no recognition.
"It's behavioral," she said. "Not structural. We fixed the language of the system, but not the habits of the people using it."

Ethan opened his laptop. "There's a pattern in the Memory Lake," he said. The room turned toward him.

"What kind of pattern?" Daniel asked.

Ethan clicked into a time-series view. "Look here. The definitions aren't drifting—but the inputs feeding those definitions were drifting."

He pointed to subtle changes in when transactions were closed, how certain categories were assigned, and how edge cases were handled by different teams.

Same definitions.

Same logic.

Different behaviors.

"The Memory Lake ingests the same meaning," Ethan said, "but the organization isn't producing consistent movement. We need to govern not only the definitions—but the flow of data."

Daniel leaned forward.
"So you're saying we need a recorder. Something that captures changes as they happen."

"Yes," Ethan said.
"A ledger. But not for transactions. A ledger for meaning in motion."

This was the first articulation of what would become the Reconciliation Ledger.

Not a standalone system.

Not a tool upgrade.

A governance discipline.

And the company needed it before things broke again—silently.

The Birth of the Reconciliation Ledger[10]

It was not a blockchain solution.

It was not automation disguised as governance.

It was, fundamentally, a governance mechanism.

Technically, it was simple:

A timestamped log table in the Memory Lake

Fed by three sources:

– automated data-quality checks

– exceptions from operational systems

– human-submitted corrections or clarifications

Supported by a lightweight orchestration workflow, Azure Data Factory

---

[10] **Reconciliation Ledger**—A structured, daily record of how aligned data shifts over time, capturing every correction, exception, override, and behavior-based deviation so trends can be analyzed and resolved proactively.

Deepak Rana

(ADF)[11], AWS Step Functions[12]—would have been the equivalent if DWL were on AWS)

Reviewed daily by the governance team and weekly cross-functionally

What made it powerful was not the mechanics.
It was the visibility.

For the first time, DWL could see the distinct 'flow signature'—the timing and pattern of corrections—associated with each function:
Finance corrected revenue lines earlier in the month
Operations made adjustments during cycle counts
Sales performed concessions near quarter-end
Regions introduced timing shifts during local holidays
Some teams overrode rules to fix edge cases faster

None of this meant malicious behavior.

It represented local optimization—small, well-intended choices that helped local teams succeed but created global inconsistency.

The Reconciliation Ledger exposed this reality in a way no dashboard ever had.

Priya articulated it clearly during the first review:
"This isn't about mistakes. It's about patterns. Humans shape data as much as systems do. The ledger simply reveals where behavior diverges from design."

Ethan added, "And once behavior becomes visible, we can redirect it—not punish it."

Daniel nodded. "Good. The board doesn't need perfect numbers—they need consistent, trustworthy ones."

The Reconciliation Ledger went live the next morning, becoming the first real-time governance instrument in DWL's history.

Within one week, DWL learned more about its real-time behavior than it had in the previous three years.

---

[11] **Azure Data Factory**—A cloud integration service that moves and transforms customer, order, or revenue data across systems so AWS and Azure can generate consistent business insights. For example, Azure Data Factory combined customer order data from AWS and Azure to improve margin forecasting.
[12] **AWS Step Functions**—A cloud orchestration service that coordinates customer, order, or revenue processes across AWS and Azure so margin-related workflows run reliably in sequence. for example, AWS Step Functions automated a customer order process across AWS and Azure, ensuring accurate margin calculations.

When Real-Time Corrections Come Alive

On Day-6 of the Reconciliation Ledger, the first significant anomaly emerged.

A cluster of corrections appeared in the operations feed from one European plant—dozens of adjustments timestamped within minutes of each other.
Not technical errors.
Not missing fields.
Human overrides—intentional adjustments made to keep local processes running.

Inventory adjustments had been made after an unexpected equipment breakdown. The plant manager had updated counts to prevent production planning from showing a misleading surplus.

Operationally, it made sense.
But it created a semantic contradiction—an operational correction that broke the governing meaning used by Finance and Sales.

In the Memory Lake, the inventory values shifted abruptly.
Finance flagged it as a reporting gap.
Operations flagged it as a necessary operational fix.
Sales flagged it as a potential availability risk.

Three perspectives.
One correction.
Multiple interpretations.

Ethan brought the cluster to the leadership team.
"This is exactly why the ledger matters," he said. "We're seeing the correction, but we also need to understand the operational intent behind it—why the change was made, not just what changed."

Priya added, "Before the ledger, this would have been invisible.
Finance would have called it a data integrity issue.
Operations would have considered it normal.
Sales would have discovered it only after promising product availability."

Daniel asked the key question. "So, what do we do with this visibility?"

Priya replied, "We don't stop the correction. We govern the pattern. We need better upstream rules so plant managers don't have to manually override to maintain credibility with planning."

Ethan nodded. "This points to a system design flaw—not a behavioral flaw. The correction isn't misuse; it's compensation for a missing upstream rule."

This distinction was critical.
A data issue is not always a people issue.
Most data issues begin as system design gaps.
So, the governance team documented the cluster, flagged it, and began tracing the cause:
Planning system assumed zero downtime.
No fallback rule for sudden machine stops.
No threshold-based alerting.
No automated adjustment logic.

Manual updates were the only way to prevent downstream chaos.
The Reconciliation Ledger made this visible.
The Memory Lake amplified it.
And leadership understood it.

Within weeks, a new plant-level rule was introduced:
If a machine goes offline unexpectedly for more than 45 minutes, the planning system triggers an automated validation check and updates inventory status using a predefined fallback rule—removing the need for manual overrides.

This was the first prospective fix emerging from real-time insights.
No models.
No AI.
Just better system behavior.

This single upstream rule reduced downstream reconciliations by 18%.

The Unexpected: When Behavior Shows Up Before Data

By Month Seven, the Ledger exposed something unexpected—an inconsistency not in the values themselves, but in the timing of how teams corrected those values.
A pattern in timing—not values.

Ethan noticed it while reviewing the flow signature of global teams. "Corrections from Region North cluster between 8–10 a.m., while Region South consistently logs corrections after 3 p.m."

He pulled five weeks of Ledger entries and plotted them by timestamp distribution.

The distributions formed mirrored patterns—one region resolving issues early, the other deferring them consistently.

Priya stared at the chart. "This isn't operational. It's cultural—habit rhythms, not process rules."
Region North started early.
Region South deferred issues later in the day.
Then he mapped weekends.
European teams tended to log corrections on Fridays, while several Asian teams concentrated their log corrections on Mondays.

"Behavior precedes data,' Ethan said. 'The data changes are downstream of human timing patterns.
"The ledger is showing us cultural fingerprints."

This was the first moment DWL saw:
Data does not just reflect operations.
Data reflects people.

Daniel understood the implications instantly.
"If the same rule is applied differently across regions because of habit rather than design,' he said, 'then much of our performance variability is cultural, not systemic."

Priya added, "It means our interventions must be tailored to behavior patterns—not enforced uniformly."

The Memory Lake had revealed numbers.

The Ledger had revealed behaviors.

But this—this was the first time DWL saw culture through its data.

It became the closest mirror DWL had ever held to itself—showing not errors, but habits.

And it changed the tone of leadership conversations permanently.

Reflection—What Flow Revealed
What we observed
The company's meanings had finally aligned, but its movements had not.
Local optimizations still produced global inconsistencies, revealing areas where human timing and interpretation diverged from system design.
The Reconciliation Ledger exposed these behavioral patterns as the primary source of drift.

What it means
Data governance is both structural and behavioral.
Real-time visibility reveals issues before they propagate downstream, allowing the organization to intervene early and deliberately.
The Memory Lake and Ledger demonstrated that culture leaves measurable signatures in data—and those signatures directly influence performance.

What will happen next
Embed real-time governance into daily operational routines
Use Ledger insights to prioritize upstream system design improvements, reducing the need for manual corrections.
Prepare for the next stage: aligning system behavior with human behavior so that corrections become proactive, not reactive.

Aphorism
Clarity is not static. It moves with people; discipline is learning to move with it.

Guiding Principle
Behavior reveals itself before numbers do.

Law — The Behavioral Precedence Law
Behavioral drift precedes and predicts data drift.

Equation
$D_{data} = alpha \times D_{behavior}$

Technical Explanation
Data drift D_{data} is linearly proportional to behavioral drift $D_{behavior}$.
When human behavior changes, the data eventually reflects it—even if systems are temporarily slow to show it.

Symbols
D_{data}: Change or drift observed in metrics and datasets.
$D_{behavior}$: Change or drift in how people act (process steps, timing, workarounds).
alpha: Positive proportionality constant linking behavioral change to data change.

PART IV—THE ALIGNMENT

The company stabilizes its systems so meaning cannot drift.
The work shifts from fixing the past to preventing recurrence.

CHAPTER 9—Prospective Control

The Retrospective Plateau

By Month Eight, DWL had accomplished something that once seemed impossible: it had reduced daily confusion.

Definitions were aligned.

The Memory Lake was stable.

The Reconciliation Ledger was in use.

Leadership could finally look at the same numbers without debating what they meant

But something else happened—quietly, almost invisibly.

The rate of improvement slowed.

Not because the teams were complacent, but because the organization had absorbed everything it could through retrospective work.

They had:
- cleaned historical backlogs
- corrected silent drifts
- flagged inconsistent patterns
- cataloged fifty core terms
- stabilized ingestion into the Memory Lake
- captured overrides and exceptions in the Reconciliation Ledger
- identified timing patterns and cultural signatures

The foundation was real.

The discipline was real.

But some problems kept coming back.

Not in the same volume.

Not with the same randomness.

But enough that the leadership team began sensing the same pattern the Memory Lake had already revealed:

Retrospective correction can stabilize the past.

Only prospective control can shape the future.

A Meeting That Marked the Plateau

The realization crystallized in a governance review meeting that Daniel convened late in Month Seven.

Around the table sat:
– Daniel
– Elena
– Aaron
– Maya
– Priya
– Ethan

No one needed to be reminded why they were there.

The ledger's last report had shown something important:
– Errors were decreasing
– Drift was decreasing
– Exceptions were clustering
– Manual overrides were becoming predictable
– Cultural timing patterns were stable

But patterns were not disappearing.
They were revealing structure.

Daniel began with the simplest question in the room. "Why," he asked, "are we correcting the same issues month after month?"

Aaron looked at the ledger summary. "It's because the triggers haven't changed. The ledger shows the symptom, not the cause."

Elena added, "And some corrections originate from the way data enters our systems. We clean it later, but that only patches the surface."

Maya tapped the table. "My teams adjust concessions because the Customer Relationship Management (CRM)[13] doesn't enforce the rules upstream. Unless the system stops them, they'll keep overriding to hit deals."

Priya nodded. "This is what we expected. Retrospective visibility is necessary. But without prospective controls—rules, validations, guardrails—nothing upstream changes."

Ethan scrolled through charts on his laptop. "Here's the plateau," he said, turning his screen so everyone could see.

The chart showed:
– Month 1 → Month 4: a steep decline in inconsistencies
– Month 4 → Month 6: a slower, steady decline
– Month 6 → Month 8: a visible flatline

---

[13] **Customer Relationship Management (CRM)**—An Enterprise customer relationship intake, tracking, and monitoring software

"We've reached the limit of backward-looking influence," Ethan said. "It's time to influence the data at its point of origin."

Daniel leaned back.
"And this is where most transformations fail," he said quietly.
"We have to move upstream before people lose patience."

Priya asked Ethan to explain "Why the Plateau Occurs?"

Ethan summarized what the Memory Lake and Ledger had shown:
1. Retrospective correction improves accuracy—but only after the fact.
    Pipelines can:
    –   clean data
    –   harmonize fields
    –   map missing values
    –   standardize formats
    –   resolve sequence errors
    But they cannot correct the behavior or system logic that created the inconsistencies.

2. Retrospective work becomes exponentially harder over time.
    As the company grows:
    –   more transactions
    –   more regions
    –   more exceptions
    –   more local adjustments
    This makes cleaning "after the fact" increasingly expensive and fragile.

3. Pipeline transformations introduce technical debt.
    The more fixing the pipeline must do downstream, the more risk accumulates upstream.

4. True stability requires prospective control.
    Meaning:
    –   ERP rules must prevent invalid entries.
    –   CRM must enforce agreed definitions.
    –   Planning systems must treat downtime correctly.
    –   APIs must reject malformed payloads.
    –   Master data rules must eliminate ambiguity.

Retrospective correction is a bridge, not a destination.

The Psychological Turning Point

The room grew quiet.

Daniel summarized what everyone already knew:
"We're good at fixing yesterday.
We're not yet good at shaping tomorrow."

Priya added, "We've built transparency. Now we need discipline."

Aaron leaned forward. "And discipline needs design. I can't expect plant managers to follow rules we haven't encoded in the system."

Elena adjusted her glasses. "Finance can't keep absorbing reclassifications. We need controls at the point of entry."

Maya said, "And Sales can't be the team compensating for upstream ambiguity."

All eyes turned to Ethan.

He said the sentence that became the anchor:
"We've reached the limit of correction.
It's time to prevent."

The Three Phases of Prospective Control

Ethan outlined the path ahead:

Phase 1—Validate at Entry
  Guardrails in transactional systems to prevent invalid or ambiguous data:
  - mandatory fields
  - dropdowns not free text
  - value checks
  - business-rule validations
  - format & consistency checks
  - exception alerts

Phase 2—Normalize in Flow
  Ensure data stays consistent as it moves:
  - upstream harmonization rules
  - mapping tables
  - automated transformations near source
  - region-specific adjustments encoded centrally

Phase 3—Protect Against Drift
Ensure definitions cannot silently mutate again without detection and formal approval:
- metadata versioning
- rule enforcement
- automated detection of definition deviation
- governance sign-off for changes

The Leadership Decision

Daniel made it explicit. "Retrospective correction was Stage One. Prospective control is Stage Two."

He paused. "And the moment we get this right; we'll finally be ready for intelligence—not sooner."

Elena nodded. "Agreed."

Aaron added, "Let's engineer the rules so people don't have to work around the system."

Maya said, "Sales needs this. Ambiguity is exhausting."

Priya said, "Governance will anchor it. But upstream teams must own it."

Ethan closed his laptop. "We'll start with the highest-volume, highest-variance objects. And we'll build from there."

Daniel concluded: "Let's build the guardrails—not the walls."

It was the first time leadership understood what **'Prospective control"** was really about.
Not technology.
Not definitions.
Not dashboards.
But momentum—and direction.

Retrospective work had exhausted its value.

Prospective control was the next frontier.

And DWL was finally ready.

The First Prospective Controls (Upstream Begins)

From Month Eight onwards, the shift from retrospective correction to prospective control began quietly—without announcements, dashboards, or fanfare.

It started with three data objects, selected because they created the most cross-functional friction across DWL:
1. Order
2. Backlog
3. Asset Uptime

These objects flowed through virtually every system—Enterprise Resource Planning (ERP)[14], CRM, planning, manufacturing—and produced the widest discrepancies in the Reconciliation Ledger.

Ethan referred to them as the 'First Three' because they behaved like amplifiers:
if they were misaligned, all downstream reports were misaligned.

The leadership team approved the order unanimously.

Why Order, Backlog, and Asset Uptime Came First

Priya explained the selection simply:
— Order was the root of revenue, backlog, customer promises, and planning.
— Backlog drove supply chain commitments and forecast accuracy.
— Asset Uptime was the heart of operations and cost variance.
Fixing these three would stabilize roughly 40% of downstream corrections.

Fixing them retrospectively was no longer enough.

To move forward, DWL needed entry-point control—the technical equivalent of closing a door before the wind disrupts the entire building.

1.   Order—The First Prospective Guardrails
In Month Eight, Sales and Operations met with IT to redesign how orders were created and modified.

This was the first time DWL used the Reconciliation Ledger as a design tool rather than a diagnostic one.

---

[14] **Enterprise Resource Planning (ERP)**—An Enterprise Transactional Software for core business processes
Deepak Rana

Ethan presented what had been learned from the last six months:
- 34% of order corrections originated from missing mandatory fields.
- 22% came from ambiguous status transitions (e.g., "pending" → "confirmed" used inconsistently).
- 18% were due to free-text inputs.
- 11% came from region-specific variations.

He then outlined "Prospective Control #1": Mandatory field enforcement. A mandatory field is not just a required input; it is a constraint that ensures the same meaning is captured every time the field is used.

It is a constraint that ensures the same meaning is captured every time the field is used.

Implemented in CRM + ERP:
- Product ID
- Customer ID
- Order Type
- Expected Ship Date
- Status using controlled vocabulary
- Reason Code for changes

"No more blank values," Maya said.
"No more free text," Priya added.
"No more two people using the same field to mean different things," Ethan concluded.

The rule went live in less than two weeks.

Error rates dropped by 14% within a single month.

It was the organization's first experience of true future-proofing.

2.   Backlog—Correcting the Flow Before It Breaks
This metric had long been one of DWL's most politically charged areas because each function interpreted it differently.

Finance saw backlog as future revenue.
Sales saw it as customer promise.
Operations saw it as a planning constraint.

The Reconciliation Ledger made something very clear:

Seventy percent of backlog drift wasn't caused by data errors—it stemmed from timing misalignment.

Different regions posted backlog at different points in the day or different stages in the process, creating inconsistent snapshots.

Ethan proposed Prospective Control #2: Time-Window Validation.
A rule that ensures backlog is captured and updated within a standardized time window across regions and systems.

Technically, it involved:
– A timestamp check in the planning system
– A synchronization job in Azure Data Factory
– An alert if a timestamp drifted outside the 3-hour global window
– A rule that delayed downstream pipeline consumption until values were synchronized

Business effect: Finance, Sales, and Operations finally saw backlog at the same point in time, not at different life-cycle moments.

"We stopped arguing about the number," Elena said later.
"And started discussing what to do about it."

3.   Asset Uptime—The First Intelligent Rule
Asset Uptime was the hardest object to stabilize because it was a real-world measurement, not just data.

Machine stops, slowdowns, and manual entries introduced natural variability.

The Reconciliation Ledger showed:
– Manual overrides during machine breakdowns
– Differing interpretations of "partial downtime"
– Operators logging downtime after shifts instead of during
– No automatic alerts for unplanned stops under 45 minutes

Aaron summarized the insight:
"We don't have a data-quality problem.
We have a timing problem."

Thus, came Prospective Control #3: Automated Downtime Trigger
A rule inside the plant system that fired whenever a machine stopped for more than 45 minutes.

Technical aspects:
– A heartbeat signal from machine sensors
– Edge device detection (standard in mid-sized manufacturing assets)

- A planning-system Application Programming Interface (API)[15] to update the status
- A validation job in the Memory Lake to catch outliers
- A local operator prompt requiring confirmation ("Yes/No")

Business impact:
Manual overrides dropped by 60% in the first cycle.
Planning accuracy improved by 9%.
Asset reliability reporting became uniform for the first time.

Ethan's Realizes, Prospective Change Requires Social Engineering

By Month Nine, Ethan recognized something important:

Prospective control is as much social as it is technical.

When controls were introduced gently—with shared rationale, clean UI, and clear benefit—teams embraced them.

When controls were introduced abruptly—especially with rigid constraints—teams resisted.

During a workshop with the European Sales team, one manager said: "If the system forces a rule that doesn't match customer reality, the rule becomes the enemy."

Ethan realized his job was no longer analysis.
It was translation—between algorithms and human behavior.

And Daniel noticed.
"You're not just fixing data," Daniel told him.
"You're learning how the company behaves. That's what will make you indispensable."

It was the first indication that Ethan's quiet awakening was becoming a critical leadership pillar.

The First Real Test: When a Control Failed

Two weeks after the first guardrails were introduced, one rule broke.

A mandatory field in CRM had been configured incorrectly.

---

[15] **Application Programming Interface (API)**—A formal set of rules that lets AWS and Azure systems share customer, order, or revenue data securely for margin analysis and business applications. for example, An API connected AWS sales data with Azure analytics to calculate customer revenue margins in real time.

DEEPAK RANA

It blocked order creation entirely for four hours.

Maya's team alerted Priya.

Within minutes, the issue escalated.

Yet something unexpected happened:
No one blamed the team.
No one blamed IT.
No one blamed governance.

Instead, the leadership team said:
"This is why we test upstream.
This is why we go slow."

Prospective control, the organization realized, was not a switch.
It was a process.
A negotiation between rules, systems, and human behavior.

And it was working.

Engineering Stability (Managing System Behavior Changes)

The three prospective controls established guardrails. Engineering stability focused on changing the behavior of the systems themselves—not merely correcting what systems captured, but shaping what they allowed.

At DWL, this phase had a name only a few people used out loud:
System Behavior Engineering[16].
It meant building rules, checks, locks, and flows so that meaning could not drift—not quietly, not accidentally, not because someone "interpreted a process differently."

This was the moment governance stopped being meetings
and started becoming machinery.

The First Real Pipeline Rewrite—Technical Turning Point

For the first eight months, the Memory Lake had been supported primarily through retrospective harmonization.
ADF pipelines cleaned, aligned, and lineage-stamped every data object before landing it in the curated zones.

---

[16] **System Behavior Engineering**—designs and manages how systems respond to events and data flows to ensure predictable outcomes, formally defined as the discipline of modeling, validating, and automating system interactions across distributed architectures; for example, Azure Event Grid and AWS EventBridge were engineered to trigger backlog updates automatically when a customer order status changed, ensuring asset records stayed synchronized.

But now, the engineering team shifted to a behavior-first model.

Goal: If an upstream system produced invalid or ambiguous data, the pipeline would flag it, block it, or request human confirmation—rather than quietly cleaning it downstream.

This was a major cultural shift.

The New Pipeline Flow—implemented across the Order, Backlog, and Asset Uptime data objects:
1. Validate input
   - Does the field conform to controlled vocabulary?
   - Are mandatory fields populated?
   - Are timestamp windows consistent?
   - Does the value violate a domain rule? (e.g., negative lead time)
2. Assign a severity
   - Low (auto-correctable)
   - Moderate (requires human confirmation)
   - High (blocks downstream ingestion)
3. Trigger behavior
   - Auto-correct based on known rules
   - Ask for confirmation from the owning team
   - Quarantine the record in a validation layer
   - Raise an alert to data stewards
   - Block the load and notify the system owner
4. Apply lineage tags
   - Validation logic used
   - Rule applied
   - Timestamp of action
   - User or service responsible

This allowed the Memory Lake to evolve from a 'clean room' into a system of accountability.

It also opened the hardest distinction: the difference between fixing data and fixing behavior.

Stability Through Controlled Vocabularies—Technical Foundation)

Meaning drift often originates in free-text fields.

Historically, DWL had allowed teams to describe operational events in free text:
   - "Cancelled," "Canceled," "Cxl," "CNX,"

- "Delayed," "Late," "Hold,"
- "Under maintenance," "Stopped," "Off-line," "Downtime,"
- "Shipped," "Sent," "Fulfilled"

Five versions of the same truth.
Ten versions of the same meaning.

So DWL introduced Controlled Vocabularies—a structured list of allowed terms that enforced semantic consistency across systems.

What changed technically
In CRM: a dropdown replaced free-text "order status" fields.
In ERP: a new code set replaced manual entries for "reason of delay."
In Planning: scenario labels were standardized (no more "alt-forecast," "rough plan," "master plan v2").
In Data Pipelines: a mapping function validated terms and flagged unexpected values.

What changed culturally
People stopped "naming" their own interpretations.
They started selecting from shared meaning.

This was the first time DWL felt stability not just in pipelines, but in language itself.

The First Automated Feedback Loop—Real-time corrections Begin

Prior to Month Nine, DWL's corrections were all backward-looking.
Now, through a combination of:
- Memory Lake lineage tags
- ADF validation logs
- Service alerts
- CRM and ERP metadata
- A small monitoring module built in Databricks[17] (lightweight job, no ML runtime required)
…the organization built its first real-time correction loop.

What this loop did:
Whenever an upstream system created a record that violated a rule:
1. Memory Lake blocked ingestion
2. The pipeline pushed an alert to the steward
3. The steward received a suggested correction

---

[17] **Databricks**—A cloud-based analytics platform available on AWS and Azure that unifies data engineering, machine learning, and business intelligence to process customer, order, or revenue data for margin insights. Example " Azure Databricks analyzed customer orders while Databricks on AWS processed revenue streams, together improving margin forecasts.

4. The steward approved or rejected
5. The system stored lineage and updated the lake
6. The originating team was notified of pattern frequency

This loop allowed DWL to do something it had never been able to do: prevent downstream inconsistencies before they propagated.

For Aaron, this was the first moment he believed technology could meaningfully reduce plant-level firefighting.
"It tells me the mistake before it burns the schedule," he said.

Stability Metrics Become Leadership Metrics

The "Governance Backbone" had established owners.
The learning from Pilot had revealed blind spots.
The "Memory Lake" had exposed timelines and behavior.

Now, Daniel introduced something bolder:
Stability Metrics—a small set of measures that leaders were responsible for:
– Definition Adherence Rate—Percentage (%) of data records using correct term definitions
– Upstream Compliance Rate—How often systems follow mandatory fields, vocabularies, and logic
– Override Frequency—How often humans override a system rule
– Pipeline Block Rate—Signals of systemic issues
– Semantic Drift Alerts—Number of meaning shifts detected across functions

These were not IT metrics.
They were leadership metrics.

Finance owned meaning in financial terms.
Operations owned meaning in operational terms.
Sales owned meaning in customer terms.

And for the first time, executives were accountable for stabilizing the system behavior.

This was the subtle arrival of the machine-governed organization."

Not AI.
Not automation.
Not algorithms.

But rules that reduced confusion.

From Analyst to System Interpreter—Ethan's Deepening Role

Ethan noticed a striking pattern in Month Ten:
When a new prospective rule was introduced, the biggest point of failure was not the system.

It was the interpretation of the rule's purpose.

So, he began running weekly sessions called:
"How the System Thinks"

They weren't technical training.
They weren't governance reviews.

They were translation sessions.
He explained:
– Why the pipeline blocked a load
– Why a field needed to be mandatory
– Why controlled vocabulary mattered
– Why the Memory Lake used lineage
– Why the timing windows needed consistency
– Why meaning had to be consistent before AI arrived

People started asking for the sessions.

Not because they enjoyed governance—but because they finally understood the "why."

Priya noticed something else:
When Ethan explained system behavior, people changed their own behavior.

It was the first time she saw him as a bridge
between technical feasibility
and organizational psychology.

And that changed the momentum of the entire program.

The First Prospective Success Story

In the Month Ten, backlog stabilization hit a milestone:

Variance across three regions dropped from 12% to 1.6%.
Not because someone reconciled faster.
Not because Ethan built a better model.
Not because Finance "forced alignment."

Not because Operations "finally followed process."

But because:
– CRM standardized order capture
– ERP validated lead times
– Planning synchronized timestamps
– Memory Lake rejected misaligned values
– ADF pipelines enforced logic
– Regional teams understood the rationale
– Leaders owned the outcome

This was the first measurable evidence that prospective control worked.

Daniel summarized it best:
"We're not fixing numbers anymore.
We're fixing causes."

The Alignment—Organizational Behavior → System Behavior Integration

By Month Eleven, DWL finally reached the stage where behavior, not data, told the true story of the enterprise.
Not because people changed first—but because the system forced transparency in ways people never had to confront before.

The Memory Lake, the upstream rules, and the reconciliation guardrails began revealing something human: patterns of behavior that had always existed but were never visible.

Real Behavior Emerges Through Real Data Fields

The first major shift wasn't in the systems.
It was in how people interacted with ten of the fifty defined terms.

Example A— "Order Date"
In many regions, Sales had used "order date" to mean:
– date of customer commitment
– date internal approval was secured
– date a rep entered the order
– date pricing was finalized

Now, the upstream rule required it to be:
"The first binding commercial commitment date agreed with the customer."
When this rule activated, the system began blocking orders where:
– the commitment date was after shipment date

- the commitment date was missing
- the commitment date was changed multiple times
- the commitment date occurred after delivery

No one expected how revealing this would be.
It showed:
- which teams delayed entering orders
- which reps "backdated" commitments
- which regions avoided lead-time metrics
- which customers negotiated aggressively

This wasn't a data problem.
It was a behavioral pattern DWL could never see before.

Example B— "Lead Time"
Lead time had been one of the biggest sources of friction for years.
- Operations treated lead time as: "the actual time required to produce and ship."
- Sales treated it as: "the time we tell the customer so we don't lose the deal."
- Finance treated it as: "the time used for revenue recognition modeling."

When the upstream rule required a single definition:
"The elapsed time between customer order date and confirmed ship date."
…the system instantly exposed behavior patterns:
- Plants padding lead-time buffers
- Sales shortening lead times to secure deals
- Planning teams adjusting lead times silently in spreadsheets
- Finance using static lead times never validated against operations

Within two weeks:
- 17% of orders triggered lead-time inconsistencies
- 9% were corrected only after escalation
- 4 plants emerged as recurring variance hotspots
People were startled.
The system was holding a mirror up to the company.

Example C—"Active Customer"
This term had been debated for years.
With one unified definition:
"A customer with
(a) at least one order in the past 12 months or
(b) an open order scheduled for fulfillment."

…Memory Lake surfaced:
- Duplicate accounts with slightly different names
- Customers marked "active" in CRM but dormant in ERP
- Regions inflating customer counts to meet targets
- Customers with no orders but open credit lines
- Ghost accounts created for "testing"

The behavioral map showed that customer activity reporting had been influenced by incentives, not reality.

Daniel's reaction summarized the moment:
"We're not losing customers.
We're losing definitions."

Organizational Assumptions Become System Truth

With prospective controls live, DWL no longer had the luxury of ignoring misalignment.

The system made assumptions visible immediately.

Finance Assumption Revealed
"Operations will fix lead times downstream."
But upstream rules now blocked orders with mismatched dates.
Operations could no longer "fix them later."

Operations Assumption Revealed
"Planning will smooth out demand fluctuations."
But forecast accuracy rules flagged inconsistencies before planning even saw the inputs.

Sales Assumption Revealed
"Finance will adjust revenue anyway."
Revenue recognition rules refused to advance if customer commitments didn't align.

Planning Assumption Revealed
"Sales data is mostly clean."
Memory Lake lineage showed 28% of forecast line items came from inconsistent customer identifiers.

Each assumption became visible because the system no longer bent to accommodate them.

Priya's Integration Workshops Become a Turning Point

Priya realized something important:
People didn't resist rules.
They resisted rules they didn't understand.

So, she launched System Integration Workshops—and this was where the real transformation began.

Each workshop followed a simple, repeatable structure:

1.   Scenario Reconstruction (Real incident walkthrough)
Case: Discount approval breakdown
  —   Customer: Global Partners
  —   Sales entered a 17% discount
  —   Policy allowed a maximum of 12% without CFO approval
  —   System blocked the order
  —   Sales attempted to override
  —   Block triggered an escalation
  —   Customer delivery was delayed 24 hours

In the workshop, Priya broke it down:
1.   Sales assumed Finance would "adjust margin later."
2.   Finance assumed Operations would absorb the variance.
3.   Operations assumed Planning had built the forecast with correct pricing.
4.   Planning assumed Sales had honored discount rules.
5.   The system exposed that all four assumptions were wrong.

For the first time, people saw how one field—Discount Percentage—affected:
  —   revenue
  —   margin
  —   forecast
  —   production sequencing
  —   customer satisfaction
  —   month-end reconciliations

Without shaming anyone, Priya simply showed the consequence chain.
It changed the tone of every workshop that followed.

2.   Behavior Mapping—Ethan's behavioral telemetry
Ethan used Memory Lake lineage to surface behavioral signals:
  —   22% of 'urgent order requests' were created by Sales reps with consistently late approvals
  —   31% of lead-time overrides originated from operations managers

bypassing planning coordination
- 19% of forecast updates were made within 48 hours of cycle close
- 14% of customer accounts were manually updated by five individuals
- 61% of exceptions were concentrated in three regions

The message was subtle but unmistakable:
"We are all contributing to the friction."

People didn't dispute the data.
They recognized themselves in it.

3. Integration Commitments—behavior → system alignment
Each team agreed to one or two commitments:
- Sales: "No order enters the system without validated customer ID."
- Operations: "Lead-time updates will not be backdated."
- Finance: "Margin adjustments require documented justification."
- Planning: "Forecast revisions after cycle close require executive approval."
- Governance: "Definition drift alerts must be reviewed weekly."

These weren't slogans.
They were behavioral contracts—agreements that turned governance into culture.

Emerging Predictive Behavior (Without AI)

By Month Eleven, Ethan noticed the most surprising shift yet: system behavior was anticipating human behavior.

Not through machine learning—but through the accumulation of behavioral signals.
Examples:
- Spikes in last-minute forecast edits predicted reporting delays
- Increased lead-time overrides predicted production bottlenecks
- Discount exceptions predicted margin reconciliation spikes
- Customer record updates predicted duplicate account growth
- Timestamp anomalies predicted inventory write-down risks

Priya described it simply:
"The system is teaching us about ourselves."

It wasn't intelligence.
It was organizational reflection made visible for the first time.

This was the beginning of alignment—the moment when system maturity and human maturity finally intersected.

Daniel's Realization: The System Is Becoming the Organization's Structural Memory

When Daniel reviewed the integration reports, he saw something deeper:

DWL's systems were no longer separate tools.
They had become structural memory—a living record of how the company actually behaved.

He summarized the shift in one line:
"People used to correct the system.
Now the system is correcting us."

This wasn't automation.

This was evolution:
— chaos → structure
— structure → discipline
— discipline → learning
— learning → prevention
— prevention → trust
— trust → intelligence

DWL had crossed a threshold—not just stabilizing data, but stabilizing meaning, behavior, and governance.

Reflection—Prospective Control and Behavior

What we observed

The shift from retrospective cleaning to prospective control created transparency around behaviors—discounts, lead times, customer data, forecast edits—that had previously been hidden by manual workarounds and downstream corrections.

What it means

Prospective controls do more than clean data. They surface assumptions, reveal systemic habits, and expose friction that originates in behavior—not in technology. The Memory Lake and upstream rules created a structural mirror that finally showed DWL how people actually interacted with processes.

What will happen next

Scale prospective controls to the next set of high-impact terms—focusing on order, customer, margin, backlog, and forecast fields.
Launch behavior—system alignment sessions in each region.
Prepare the organization for next evolution: Governance as Code—aligning system behavior with organizational intent.

Aphorism

Systems become wise when people behave consistently. People behave consistently when systems reflect what they value.

Guiding Principle

Integration is the moment when assumptions collide.

Law—The Assumption Collision Law

The quality of integration is limited by the clarity of underlying assumptions.

Equation

$Q_{integration} = min(A_1, A_2, ..., A_n)$

Technical Explanation

Integration quality $Q_{integration}$ is capped by the least clear assumption among $A_1...A_n$. One weak or hidden assumption can limit the effectiveness of an entire integration effort.

Symbols

$Q_{integration}$: Overall quality of integration across systems/functions.
A_i: Clarity of assumption set i (for each function or system), on a comparable scale.
$min(\cdot)$: The minimum operator, selecting the lowest value among the assumptions.

CHAPTER 10—Echoes of Order

The Restart

DWL's leadership approached the Month Eleven review with a quiet caution that hadn't existed earlier in the journey. The early victories of the Language of Truth, the Memory Lake, and the Reconciliation Ledger had stabilized the organization in visible ways—fewer escalations, fewer late-night reconciliations, fewer contradictions leaking into leadership meetings.

But stabilization wasn't the same as progress.

The board had approved a pause on new AI work back in the first month. It wasn't a symbolic gesture—it was a hard stop. No model development, no algorithm experiments, no exploratory initiatives in the background. Earn the right to build AI had become the quiet mantra guiding everything DWL did.

Now, for the first time since the pause, Daniel Shaw believed the organization could begin again—slowly, deliberately, and this time, from a foundation that would not collapse under its own contradictions again.

The restart meeting had only six people seated at the table:
1. Daniel
2. Elena
3. Aaron
4. Maya
5. Priya
6. Ethan

This wasn't because others weren't important. It was because restarting AI required alignment at the very top before anyone else touched a line of code or a dataset. DWL had learned enough in ten months to know that starting too wide, too fast, with too many hands would re-create the same fragmentation they had just spent nearly a year fixing.

Daniel opened the meeting with the clarity that had become his trademark. "We've built stability," he said. "Now we make it operational."

No slides. No charts. Just the statement.

Priya adjusted her notebook. "The Reconciliation Ledger has flattened. We're down to fewer than twelve open corrections a week. Those are mostly explainable delays, not structural problems."

Ethan added, "And the Memory Lake is finally predictable. No more silent divergences. No mid-stream schema shifts. Lineage checks are passing consistently."

Elena exhaled through her nose, not dramatically, but in that quiet way that signaled cautious relief. "Which means," she said, "we can trust the baselines."

For the first time since the crisis began, "trust" wasn't used as a metaphor. It was measurable. Variance across the ten foundational objects—Order, Shipment, Invoice, Asset, Customer, Plant Event, Forecast, Supplier Commit, Return, and Contract—had dropped from double digits in the second month to under 2% by eleventh month.

That wasn't perfection.
But it was alignment.

Daniel leaned forward. "If we restart AI now, we do it with discipline.
No parallel projects.
No scattered experiments.
One program.
One architecture.
One governance path."

Priya nodded once. "Athena."

The name had existed for months—quietly, in the background—waiting for the moment when the foundation could support the intelligence built on top of it.

Ethan spoke carefully. "If Athena is the umbrella, then we need to define the starting point. One model, not many."

Aaron added, "And something operational. Something that matters."

The group fell silent for a moment. Not because they lacked ideas, but because restarting AI wasn't about ambition. It was about making sure the first model they built—after eleven months of discipline—did not set off a chain reaction of drift.

Daniel looked at Elena. "Where do we feel the pain most clearly?"

Elena didn't hesitate. "Forecast variance. It's still too wide. We've solved the meaning problem, but we haven't solved the anticipation problem."

Aaron agreed. "And demand swings hit my plants before finance has modeled them. If Athena can help narrow that gap, it pays for itself."

Ethan added the analytical layer. "We already have clean data for Orders, Shipments, Forecast Adjustments, and Plant Events. We're closer to a stable time series than anything else."

Priya summarized. "So, the first model isn't a moonshot. It's a calibration model—designed to help us see the future a little earlier, not to automate anything."

Daniel nodded. "Exactly. A model that helps us understand ourselves, not one that replaces decisions."

It was the first time in the entire project that AI was positioned not as a disruptor, but as an extension of organizational awareness.

And that was the beginning of Athena.

"Before we start," Priya said, "I want one clarification on the record."

Everyone turned.

"No model launches," she said, "without a pre-defined source of truth for every field it touches. That means the Language of Truth controls the semantics, the Memory Lake controls the data, and the Reconciliation Ledger controls the history. No exceptions."

No one disagreed.

It was no longer a project rule.
It had become culture.

Daniel closed the meeting with a sentence simple enough to become a milestone.
"We begin again. And this time, we begin aligned."

The next phase of DWL's transformation would not be about cleaning the past, but shaping the future—deliberately, predictably, and with an intelligence the company had finally earned.

The Moment DWL Began to Stabilize Its Future

By the Month Twelve, DWL had entered a territory that few organizations reach.

Misalignment was no longer invisible.
Definitions were governed.
Retrospective corrections were slowing down—not because effort decreased, but because upstream understanding increased.

Yet something else was happening, quietly but unmistakably.

The systems were beginning to behave.

Not perfectly.
Not consistently.
Not magically.

But predictably.

And that was enough to change the tone of the entire transformation.

Systems Begin to Behave

The first indicator of change was not a dashboard.
It was the absence of panic.

For months, Kara's 7 a.m. reconciliation review had been a ritual of discovery:
- drift in a supplier maturity score
- a subtle shift in planned vs actual window
- an unannounced change in sales pipeline stage logic
- a quiet override of an allocation rule

But midway through the Twelfth Month, something different began to appear in her morning routine:

Nothing had changed.
Because nothing had been allowed to change.

The Memory Lake's "Truth Layer" and the Reconciliation Ledger were now acting as early-warning sensors.
Not for data quality—those issues were old news—but for semantic behavior.

One morning, she pointed it out to Ethan.
"There haven't been new anomalies in two weeks."

He looked at the audit logs. She was right.

The ERP field "Customer Type" had stopped shifting unannounced.
The CRM field "Opportunity Stage" had remained consistent across regions.
"Order Date" was no longer moving between functions.
"Active Customer" had no new flavors.

Ethan asked, "Why now?"

Kara replied, "Because people stopped guessing. The rules are real now."

The change wasn't dramatic. It wasn't loud.
But it was unmistakable.

Systems that once reflected internal confusion were now reflecting internal memory.

Humans Begin to Adjust

By the Month Twelve, a second shift emerged.
This time, it wasn't the systems.

It was the people.

Meetings that once spiraled into definitional chaos became shorter.
Reconciliation calls moved from tri-weekly to weekly, then to bi-weekly.
Pushback from functions softened—not because they were compliant, but because they understood why alignment mattered.

The emotional cadence of DWL was stabilizing.

During one cross-functional meeting, a local controller challenged the definition of "Revenue Commitment Date."

Instead of debating, the group did something new:
They pulled up the governing definition, the lineage, the owner, and the rationale—within 30 seconds.

The debate dissolved.
The decision held.

Priya whispered to Ethan afterward, "They're not asking for their own definitions anymore. They're asking for the company's definitions."

This was more than adoption.
It was identity formation.

Another example surfaced in operations.

A plant manager reviewed the "On-Time Performance" definition and realized his team had been logging planned completion dates incorrectly—not maliciously, but because the legacy field label misled them.

After training, they corrected the behavior themselves without escalation.

This was the cultural milestone Daniel had anticipated:

Alignment was no longer something leaders demanded.
It was something people protected.

DWL wasn't just behaving differently.
It was thinking differently.

The First Signs of Harmony

By the Month Thirteen, DWL reached a point few organizations ever reach.

Not harmony in the idealistic sense.
Harmony in the structural sense.

Semantic drift had slowed.
Data conflicts were shrinking.
Upstream rules were stabilizing.
And for the first time—DWL could listen to itself.

This was the first capability that made "Measurement of Wisdom" possible.

Ethan and Mara gathered evidence from the Memory Lake:
–   lineage depth increased from 30% to 65%
–   definitional conflicts dropped by 40%
–   lag between upstream change and downstream detection fell from 5 days to under 24 hours
–   prospective rule adoption reached 72% across four major systems

During a leadership review, Aaron remarked:
"For the first time, I trust that the number I'm seeing is the number they're seeing."

Elena added, "For the first time, the P&L discussion wasn't a debate."

Maya concluded, "We're not aligned everywhere. But the parts that matter most are aligning."

These weren't exaggerations.
They were objective signs of systemic stabilization.

DWL was not fully aligned. But it was no longer drifting.

The company had entered the plateau of predictable behavior—the essential precursor to model transparency and machine-supported judgment.

And it set the stage for the next breakthrough.

Governance as Code (GoC)[18]: When Rules Become System Behavior

The decision to restart Athena had forced DWL to confront a truth almost no one had articulated in the early months:
Alignment is not a workshop output—it is an operating system.

The Language of Truth had given them shared meaning.
The Memory Lake had made their past visible.
The Reconciliation Ledger had kept their corrections honest.

But none of those things could scale unless the rules they agreed on stopped living only in conversations and started living in code.
This was the month DWL learned what that actually meant.

During the thirteenth month, a working session took place in a smaller room—no slides, only whiteboards and a few laptops. Present were: Daniel, Priya, Ethan, a senior architect named Mara Sengupta, and a configuration specialist from ERP team named João Ramos.

The goal was simple:
Turn governance from people remembering rules into systems enforcing rules.

Mara began. "If Athena is going to use consistent semantics, we have to embed those semantics where data is born—not just where it lands."

Daniel nodded. "You're saying what we agree in meetings must be implemented upstream."

"Exactly." Mara tapped the whiteboard. "This is what I call governance as code. We translate our definitions into validations, field logic, and guardrails at the point of entry. Not after the fact."

---

[18] **Governance as Code (GaC)**—The practice of expressing data and compliance rules in machine-readable scripts so they can be tested, version-controlled, and deployed like software.

Ethan added, "Otherwise the lake becomes a correction engine. And that won't scale."

João drew a small diagram of a master data form in the ERP system. "Let's take the simplest example," he said. "Customer Status."

He listed the current states on the board:
—   Active
—   Inactive
—   Eligible
—   Ineligible
—   Limited
—   Pending
—   Internal
—   External

He circled three of them. "These were added over the years because different teams couldn't agree on what 'Active' actually meant. So, they invented flavors of 'Active.'"

Priya looked at Daniel. "Exactly what the ledger found. Eight meanings for one term."

João continued. "We proposed reducing this to three states: New, Active, Dormant. But this time the definitions are governed, and the transitions are controlled by rules—automated rules."

He began writing:
—   New → Active, only when order volume meets agreed threshold.
—   Active → Dormant, only if no activity for 180 days, not manually triggered.
—   Dormant → Active, only through a documented sales reactivation process.

Mara added another row beneath it.
—   No free-text overrides
—   No local variations
—   No hidden fields
—   All transitions logged

Ethan summarized the transformation. "We're replacing interpretive logic with explicit logic. The moment someone enters a value that doesn't conform, the system pushes back—not the analyst downstream."

Daniel leaned back and said quietly, "That's alignment in motion."

Then came the harder part.

Priya brought up the next issue. "Some of our terms don't drift because of misunderstanding—they drift because of pressure."

She flipped to a slide showing the term "Order Date."

"Sales logs it when the contract is signed.
Operations' logs it when the system creates the job.
Finance logs it when revenue is recognized."

Mara nodded. "This one can't be resolved through definitions alone. We need system typing."

She drew three fields on the board:
– Order_Signed_Date
– Order_System_Date
– Order_Recognition_Date

"And then one governed field," she continued:
– Order_Truth_Date

Ethan raised an eyebrow. "Truth Date?"

"It's the one that AI models will use," she said. "The aligned definition. The one that will be consistently computed across systems."

"Where does that computation happen?" Daniel asked.

"In the pipeline," Mara said. "But only after each upstream system has standardized its own input."

Priya added, "This is governance as code. Not politics. Not preference. Behavior encoded into the system."

Daniel smiled slightly. "And if someone tries to bypass the rule?"

Mara responded without hesitation. "The system stops them."

That was the pivot.

For the first time, leadership understood that governance—when done right—was not bureaucracy but permeability. Rules people debated now became rules systems enforced.

Ethan asked the question that clarified everything:

"How long until we can embed all ten core definitions upstream?"

João answered honestly. "Three months at best, six months more realistically. We need to coordinate ERP, CRM, Planning, and Finance systems."

Priya nodded. "And that maps exactly to our timeline. We fix semantics upstream while Athena's first model is trained and validated."

Daniel concluded the meeting with a line that would mark the beginning of the organizational alignment:

"We are no longer cleaning the past. We are shaping the future. And the future is enforceable."

He paused, then added something no one expected:

"And this time, we do not drift back."

DWL had finally crossed a line—quietly, without ceremony.

Governance was no longer a project.
It had become design.

When Models Meet Rules

The moment Athena's first model touched governed data, DWL discovered something it had never experienced before: predictability.

Not prediction—predictability.
The difference mattered.

For ten months, the Memory Lake had reflected reality.
For two months, the Reconciliation Ledger had stabilized history.
For one month, Governance as Code had begun enforcing rules upstream.

Now at Month Thirteen, with everything finally aligned end-to-end, the company was ready for the first meaningful technical moment of the entire transformation:

A model and a governance system would meet—cleanly, without

contradiction.

The kickoff for Athena's first forecasting calibration model took place in the same room as the governance session. The deliberately small group now grew to include:
—   Mara (Principal Architect)
—   Ethan (Strategic Analytics)
—   Priya (Governance & Risk)
—   João (ERP configuration lead)
—   Rania (Data Engineering Lead)
—   Two planning analysts
—   One supply chain analyst
—   A finance manager for forecasting accuracy

The model was intentionally simple:

Time-series forecasting[19] with feature-based calibration[20].
No neural networks[21].
No ensembles[22].
No "AI hype."
Just disciplined, transparent modeling.

Ethan began with a simple explanation for everyone, including the non-technical leaders:
"The purpose of this model isn't to forecast better than humans. It's to detect whether the inputs we're using behave consistently with the rules we've defined."

Priya added, "In other words: the model is testing us, not the future."

It was a philosophical shock for some in the room.

For years, AI was sold as a machine that predicts.
Here, AI was being used as a machine that tests meaning.

Mara shifted the conversation to mechanics.

---

[19] **Time-Series Forecasting**—A method that uses historical data points over time to predict future trends or values. for example, Time-series forecasting projected backlog levels by analyzing monthly order patterns, helping optimize asset planning.
[20] **Feature-Based Calibration**—A technique that adjusts a model by fine-tuning specific input features to improve prediction accuracy. for example, Feature-based calibration in Azure Machine Learning and AWS SageMaker refined backlog forecasts by weighting order volume features more precisely.
[21] **Neural Networks**—A computing model inspired by the human brain that learns complex patterns from data to make predictions or classifications. for example, Neural networks in Azure Machine Learning and AWS Deep Learning AMIs identified customer purchase trends to forecast asset demand
[22] **Ensembles**—A method that combines multiple models to produce stronger, more reliable predictions than any single model alone. for example, Ensembles in Azure ML and AWS SageMaker improved accuracy of backlog estimates by merging several forecasting models

DEEPAK RANA

"We'll start with the four stabilized objects," she said, pointing to the board: Order, Shipment, Forecast Adjustment, and Plant Event.

"These four objects now meet all four governances requirements:"
1. Stable definitions
2. Upstream validation
3. Pipeline transparency
4. Ledger-corrected history

Rania added, "We will ingest these into the model training pipeline through governed streams. No raw feeds. No unaligned fields. Everything passes semantic checks before it hits model preparation."

Daniel (who attended the first half of the session only) summarized it in a way everyone understood:
"This is the first time in DWL's history that a model is learning from meaning—not from noise."

The First Technical Intersection

The first wrinkle came 30 minutes into training.

The model didn't crash.
It didn't alert.
It simply returned an unusually clean diagnostic output:

"Zero invalid semantic combinations detected."

This had never happened before.

Forecasts matched events.
Events matched orders.
Orders matched shipments.
Shipments matched timelines.

Ethan stared at the dashboard. "This is the first time I've seen the model not flag a contradiction at ingestion."

Rania responded quietly, "Because for the first time, the system didn't have contradictions to flag."

Mara leaned closer to her screen. "Look at the lineage chain. We finally have deterministic flow."

Priya asked, "Meaning?"

Mara turned the laptop so everyone could see.
"It means the model sees the same truth the humans see."

There was no applause.
No excitement.

Just the quiet comprehension of a breakthrough earned the hard way.

The First Human Test

The second wrinkle came two hours later—not technical, but behavioral.

The model generated its first calibrated forecast for a pilot region.
The accuracy improvement was small—only 1.8% better than the human-adjusted forecast.

But the deeper insight was hidden in the model's explanation:

"Forecast volatility reflects input discipline, not market behavior."

Ethan explained it in plain English:
"This isn't a forecasting variance problem. It's a process consistency problem. People make too many adjustments. The model is telling us our humans are overcorrecting."

This was the moment real AI enters a culture—not as a hero, but as a mirror.

Maya (by video) reacted with honesty: "So the model is saying my teams are making noise, not clarity."

Priya added gently, "Not intentionally. They're responding to local pressures. This is what the Ledger warned us about: psychological drift."

Daniel summarized the moment cleanly:
"The model is stable. We are not."

This didn't mean people were wrong.

It meant the organization didn't yet trust quiet data.

AI had just produced the first honest conversation about behavior, not numbers.

Model Behavior Meets System Behavior

Mara brought the room back to technical reality.
"The model is stable because upstream systems are stable. If we want forecasting to improve, we need process discipline, not model tuning."

This was a crucial distinction:
– Tuning the model would hide the organizational drift.
– Stabilizing the behavior would make the model honest.

Daniel approved one new rule:
"No manual forecast adjustments without a reason code."

Rania added the technical detail:
"We'll enforce this upstream in the planning system. No adjustments can be saved unless a reason code is entered—and reason codes will be governed."

Ethan added,
"And the model will monitor bias between reason codes and outcome accuracy."

This was the precise moment when governance as code met AI as calibration.

Models were no longer statistical tools.

They were behavioral sensors.

DWL had stepped into a new maturity layer—one where machine learning didn't replace humans…
…it disciplined them, gently, transparently, and with evidence.

The Switch from Retrospective to Prospective Stability

For more than a year, DWL had been repairing its past.
Every major insight, every reconciliation, every breakthrough had come from looking backward—discovering how definitions had drifted, where timelines didn't match, and why different functions believed different versions of the same truth.

Now, for the first time, the company was ready for stability that originated upstream rather than downstream.

This was the moment when DWL stopped fixing yesterday and started shaping tomorrow.

The First Controlled Experiment

Priya opened the governance meeting with a sentence that felt like both a warning and a milestone.

"We're entering the phase where mistakes become structural."

Not because the team lacked discipline—but because upstream corrections, once embedded, would propagate everywhere:

Planning → Financials → Operations → Memory Lake → Forecast Model → Leadership Dashboards.

João, the ERP configuration lead, brought the agenda to life.
"Today we propose the first prospective guardrail: controlling what 'Order Start Date' means across systems."

This one term had caused friction for years.
In different corners of DWL's ecosystem:
—   Sales recorded the date the customer said yes.
—   Operations recorded the date the work order was created.
—   Finance recorded the date revenue recognition began.
—   The Model used whichever version arrived first.
Three valid dates.
One invalid mess.

João explained the approach using language any leader could understand.
"We're not eliminating these dates. We're naming them correctly. And then we're choosing which one the system treats as the official 'Order Start Date' for forecasting and operational planning."

He wrote the new mapping:

Order_Start_Date → System-governed, event-driven
Order_Commercial_Date → Sales "yes" date
Order_Recognition_Date → Finance "start consumption" date

Then he circled the first.

"This is the one Athena will use. And this will be populated automatically, only when both Sales and Operations events occur. No one can type it. No one can override it. No one can modify it manually."

Daniel asked the question any CEO would.
"What happens if a region insists on a local interpretation?"

Mara answered without hesitation.
"They won't be able to save the transaction."

Silence—clean, decisive silence—followed.

For thirteen months, DWL's governance had been a combination of communication, clarification, and persuasion.

This was the first moment it became law.

Behavioral Impact Surfaces

The first test of the new rule surfaced not from technical teams, but from Sales.

A regional manager emailed Maya:
"Why is the system refusing order entry? We've always logged customer approvals immediately—we can't wait for operations to set up the internal job."

Maya brought the issue to Ethan, as she had begun doing more frequently.
"This feels like friction," she said, "but I'm not sure if it's the wrong kind."

Ethan pulled up the logs.
"The system is working as intended. They tried to set a commercial date without a corresponding internal job. In the old system, that was fine. But it created forecasting spikes."

Maya exhaled.
"So, this is the moment where governance stops being an idea and starts having teeth."

Ethan nodded.
"This is also where people learn that behavior shapes the model, not the other way around."

She didn't argue.
She didn't escalate.
She simply nodded and said:
"I'll talk to them."

That was the cultural shift DWL had been waiting for.

The First Prospective Calm

The follow-up report from Data Engineering was subtle but profound:

"Forecast volatility reduced by 0.7% in the first week."

Why?

Because upstream behavior became consistent.
Not perfect. Not universal. But consistent enough for the model to stop correcting for human variation.

For the first time, the Memory Lake wasn't just stable because the past was fixed—it was stable because people were behaving consistently in the present.

Daniel summarized the milestone in a leadership review:
"Retrospective governance taught us who we were. Prospective governance is teaching us who we are becoming."

No single change explained the shift. It was the cumulative weight of small corrections:
–   No more ambiguous order entry
–   No more free-text timeline changes
–   No more silent field interpretations
–   No more downstream fixes
–   No more contradictory calendar logic
–   No more hidden variations
–   No more local overrides

Meaning had finally started to behave the way it was defined.

And for the first time, the system behaved with it.

Ethan's Turning Point

Late one evening, after most teams had logged off, Ethan revisited the Memory Lake dashboards.

Not because he doubted the numbers—because he wanted to feel the shift.

For months, he'd watched timelines jitter, reason codes appear inconsistently, and flags light up for unexpected combinations. Even after the big cleanup, he could still feel the system "tensing" like a body recovering from injury.

But tonight, the charts were quiet.

No spikes.
No semantic conflicts.

No unexplainable gaps.
No urgent pings from the Ledger.

He whispered—not to anyone present, but to himself:

"It's finally holding."

This wasn't the victory of a model.
It was the victory of meaning.

It was the moment he realized that prospective stability wasn't the absence of noise—it was the presence of discipline.

He sent a single message to Priya.

Ethan:
The lake held today.
No corrections.
No conflicts.
I think we're ready for the next layer.

Priya replied simply:
Then tomorrow we show the organization what stability feels like.

In that moment, DWL finally crossed the invisible line between alignment and integration.

They weren't just cleaning their data or clarifying their terms.

They were building a company that behaved consistently with its own truth.

A company that could now support real AI—not as a patch, not as a dashboard, but as a partner.

The shift from retrospective → prospective stability was complete.

Reflection—Echoes of Alignment
What We Observed
Systems stopped creating silent meaning changes. Cross-functional teams began using shared definitions rather than local variations. Reconciliation reduced because it was no longer needed as often. Upstream guardrails started shaping behavior before errors were created.

DWL experienced its first measurable cultural alignment.

What It Means
Governance has transitioned from documentation → design → behavior.
Systems began enforcing clarity, not analysts.
DWL crossed the maturity threshold required for explainable AI and machine-supported judgment.
The next phase—model transparency and confidence reporting—can only happen because alignment has stabilized.

What will happen next
Introduce model interpretability: reason codes, confidence signals, traceability.
Begin designing Athena's "explainability layer."
Keep tightening upstream rules to prevent reversion.
Prepare leadership for machine-supported judgment (not automation yet).

Aphorism
Order is not silence. It is the sound of a company remembering what it agreed to mean.

Guiding Principle
Governance is most powerful when embedded, not enforced.

Law — The Embedded Governance Law
Governance implemented in systems scales faster than governance implemented in policy.

Equation
$$G_{impact} = k_5 \times S_{embedded}$$

Technical Explanation
Governance impact G_{impact} grows with the degree of system-embedded rules $S_{embedded}$.
Policies on paper matter, but rules in ERP, CRM, and workflows drive real, consistent behavior.

Symbols
G_{impact}: Effective impact of governance on day-to-day operations.

$S_{embedded}$: Extent to which rules and controls are implemented inside systems.
k_5: Positive constant converting embeddedness into realized impact.

CHAPTER 11—The Measure of Wisdom

When Insight Requires Restraint

By Month Fourteen of DWL's transformation, the company had entered a new phase—quieter on the surface, far more consequential underneath. The firefights had slowed. The reconciliations were no longer daily emergencies. The Memory Lake had stabilized into a living, traceable structure. Upstream rules were holding. The fifty governed terms had clear owners, documented meaning, and behavioral guardrails.

But now a different question surfaced, one that no system could answer on its own:
"Are we ready to let a model make a judgment before we make one ourselves?"

It was Daniel who raised it first during a weekly leadership session.
He didn't ask it rhetorically.

The room was quiet.
Not tense—contemplative.

Executives who once argued over numbers were now grappling with something deeper:
the difference between predicting and understanding.

For months, they had discussed AI as something in the future—something that would matter once they "cleaned up the present." But as systems stabilized and definitions held, the question shifted. Now they weren't debating AI's usefulness. They were debating AI's readiness—and their own.

Priya summarized it simply:
"Insight without integrity is noise.
But integrity without insight becomes a ceiling."

Everyone understood what she meant.

Why AI Could Finally Be Discussed Seriously

The Memory Lake had begun revealing not just numbers, but behavioral signatures—patterns that emerged only because:
- definitions were consistent,
- upstream rules prevented drift,
- and the Reconciliation Ledger made every deviation visible.

This meant something profound:

DWL was no longer interpreting noise. It was interpreting itself.

Ethan had been watching this shift more closely than anyone. The lake had stopped fighting them. Pipelines were flowing cleanly. The "top ten" governed terms—margin, active order, backlog, utilization, exception hours, etc.—had become stable data objects. Not perfect, but coherent enough to act as inputs to a real model.

For the first time, Ethan saw the possibility of a system where meaning didn't need to be manually defended every month.

The Fear No One Wanted to Say Aloud

Despite the progress, no one was eager to let a machine "decide" something. Even suggesting it felt premature.

Elena voiced the concern first.
"If we put a model on top of our current stability, are we reinforcing it… or are we risking it?"

She wasn't skeptical of AI; she was skeptical of overconfidence.

Aaron added another angle:
"Our operations teams are just now trusting the reconciled views.
A model that challenges their numbers on day one might set us back."

And Maya—whose function would be directly impacted by predictions—framed the human side:
"If a model tells a salesperson something their experience contradicts, we will lose credibility unless we can explain why the model thinks that way."

That was the key.

AI was not a question of capability.
It was a question of explainability.

DWL had learned the hard way that ambiguity corrodes trust.
They were not willing to repeat it with a model.

The Turning Point—The Selection Principle

Priya eventually articulated the threshold that made the room settle:
"We don't select a model because it is powerful.
We select a model because it is accountable."

The principle caught on immediately.

The model had to meet four criteria:
1. Traceable inputs—everything must tie back to a governed term.
2. Explainable factors—every inference must have visible reasoning.
3. Stable features[23]—no drifting definitions allowed.
4. Business governance—not just technical oversight, but functional accountability.

Ethan felt the weight of this.
The company wasn't asking for accuracy first—they were asking for legitimacy.

A model would need to behave like a participant in the ecosystem, not a black box.

The first viable model emerges—quietly, naturally

Once the criteria were clear, something else became obvious:
The first model had already been chosen—by the data itself.

Because DWL now had:
— clean historical records of the top ten governed terms
— stable timelines
— consistent upstream rules
— and a Memory Lake structured for sequence analysis
…the only model class that made sense as their first step was:
A time-series forecasting model
with explainable regression components.

Not glamorous.
Not cutting-edge.
Not something that would end up in a keynote.

But it was the right model because:
— it used governed data
— it was predictable in behavior
— it could be explained
— it matched the maturity of DWL
— it scaled with the business
— and it reflected the company's operational rhythms

---

[23] **Feature**—A feature is an individual piece of information (a variable) that a machine learning model uses to learn patterns and make predictions. Technically In ML, a feature is a measurable property or attribute of data that serves as an input to a model, enabling training, prediction, and analysis. for example, In AWS SageMaker Feature Store and Azure Machine Learning Feature Store, customer order frequency was stored as a feature to improve backlog forecasting accuracy.
DEEPAK RANA

This was not the moment for neural networks or deep learning[24].
This was the moment for earned prediction.

The question that moved the room

Daniel asked Ethan:
"If we build this, can you show us not just the prediction…
but the reason behind the prediction?"

Ethan nodded.
He already knew the answer.
Because the answer was yes—but only because DWL had spent months fixing meaning, not models.

Priya summarized it best:
"If the model can explain itself in our language,
then we're ready to listen."

And with that, began not just the next stage of DWL's maturity, but the next stage of its identity:

Insight with integrity.
Prediction with governance.
Wisdom built from truth.

How DWL Prepared Data for Its First Model—Feature Engineering[25] by Governance

By Month Fifteen, once DWL agreed on the type of model it was ready for, the company moved into a mode it had never operated in before: intentional preparation. Not the rushed, reactive cleanup that once consumed their quarters—but structured, governed, fully traceable readiness.

This phase had a name inside the analytics team, though it never became an official term:
Feature Engineering by Governance.

It meant something very simple and very rare in most enterprises:

---

[24] **Deep Learning**—is a type of artificial intelligence that uses layered neural networks to learn complex patterns from large amounts of data. Technically it is a subset of machine learning that employs multi-layered neural networks to automatically extract features and improve predictive accuracy in tasks involving high-volume or unstructured data. for example, In AWS Deep Learning AMIs and Azure Machine Learning, deep learning models analyzed customer order histories to forecast backlog levels and optimize asset planning.

[25] **Feature Engineering**—The process of selecting and shaping data so a model can understand it. It is the transformation of raw data elements into structured predictors optimized for model learning. for example, Turning "Exception Count" and "Forecast Variance" into aligned, standardized fields before predicting Active Order Risk Status.

DEEPAK RANA

"We don't engineer features by guessing.
We engineer them by meaning."

Step 1—The "Top Ten Stable Objects" Become Features
The Memory Lake had already stabilized ten core data objects—direct descendants of the fifty governed definitions.
These "Top Ten" were:
1. Margin (governed, reconciled)
2. Active Order
3. Active Customer
4. Backlog
5. Exception Hours
6. Asset Utilization
7. Unit Volume
8. Lead Time (standardized)
9. On-Time Rate
10. Customer Renewal Probability[26] (governed pre-AI metric)
11. Network Availability

Each of these objects was now:
– defined
– owned
– lineage-tracked
– upstream-governed
– behavior-stabilized

This allowed them to serve as inputs without the usual ambiguity that breaks most corporate models.

Ethan described it in one of the first model design meetings:
"We're not selecting features.
We're inheriting them from our governance."

For the first time, data was predictable enough to behave like a system.

Step 2—The Timeline Becomes a Model Asset
The Memory Lake's harmonized timelines allowed the team to create time-aligned features that leadership could understand intuitively:
– Lagged values (margin at t-1, t-2...)
– Seasonality indicators

---

[26] **Customer Renewal Probability**—measures how likely a customer is to continue using a product or service. Technically it a predictive metric that estimates the probability of a customer renewing a contract, subscription, or service, based on historical data and behavioral patterns. for example, In AWS SageMaker and Azure Machine Learning, models calculated customer renewal probability by analyzing backlog orders and asset usage, helping forecast future revenue stability.

DEEPAK RANA

- Cycle signatures
- Event flags (major outages, plant transitions, pricing changes)
- Stability markers (governance rules activated, upstream corrections implemented)

This was crucial.

Because in most companies, timeline misalignment introduces noise so strong that no model can make sense of it.

At DWL, however:
- order timelines aligned with shipment timelines
- shipment timelines aligned with cost recognition
- cost timelines aligned with margin recognition
- and all of them aligned with governed definitions

This allowed features to reflect not just math—but the business itself.

Step 3—Behavioral Features (Introduced Carefully and Realistically)
As Priya suggested, the model also required behavioral signals—but not psychological profiling, and not anything invasive.

Instead, DWL implemented behavioral governance signals, all strictly operational:
- Override Frequency—how often a term had been manually adjusted
- Reconciliation Delta Magnitude—size of discrepancies in the ledger
- Exception Resolution Time—how long it took to resolve operational exceptions
- Definition Challenge Count—how many times teams contested a meaning
- Data Drift Alerts—when upstream inputs changed unexpectedly
- Governance Rule Hits—number of automated rule catches in pipelines

These were organizational behavior indicators, not personal behavior indicators.

A subtle but critical difference.

Priya insisted on this boundary:
"We are monitoring the system's behavior, not the individual's behavior."

Executives appreciated this distinction.

It kept the work ethical, non-invasive, and fully aligned with governance integrity.

Step 4—Creating the First "Explainable Feature Panel"
Ethan, Kara, and the analytics team assembled a structured panel of features for model evaluation:

Group 1—Core Business Measures
Margin, backlog, active orders, unit volume

Group 2—Operational Rhythm Measures
Utilization, exception hours, lead times, on-time rate

Group 3—Behavioral System Measures
Override frequency, reconciliation delta magnitude, rule hits

Group 4—Temporal Measures
Seasonality, cycles, lagged values, event flags

Each feature included:
— definition
— lineage
— business owner
— governance checks
— data quality indicators
— transformation logic (minimized due to upstream fixes)

For the first time in DWL's history, a feature panel could be reviewed by:
— CFO
— COO
— Head of Sales
— Head of Governance
— CEO
…and everyone could understand what they were looking at.

No translation layer required.

Maya said it best:
"This is the first time I can read a model design and not feel like I'm reading another language."

Step 5—The Integrity Gate
Before a single line of model code was written, Priya instituted what became known as the Integrity Gate, a three-part review:
1.   Does every feature trace back to a governed definition?

If not, it cannot be used.
2. Can every feature be explained to the board in two sentences?
   If not, it cannot be used.
3. Would the business act on a prediction made with this feature?
   If not, it is not ethically viable.

This gate prevented complexity for the sake of complexity and forced the right intellectual discipline:
Clarity before sophistication.
Meaning before modeling.
Ownership before prediction.
A different kind of tension emerges

While the feature panel was strong, something else surfaced—a tension that leadership had never experienced before:

For the first time, business maturity constrained technical ambition.
And that was a good thing.

The model wasn't going to be:
— the most accurate possible
— the most complex
— the most "state-of-the-art"

It was going to be the most aligned.

Daniel summarized the transformation:
"We don't need a powerful model to show we've matured.
We need a responsible one to prove it."

The decision was not glamorous.
But it was wise.

And wisdom—not speed—was the point.

How the Model Learned to Explain Itself

By Month Fifteen, as DWL committed to fully building and relying on its first predictive model, Daniel made one non-negotiable demand:
"If the model can't explain its reasoning, it doesn't belong in this company."

This single requirement forced the team into a discipline most enterprises skip:
"Explainability by design, not as an afterthought."

The model wasn't allowed to be a black box.
It had to be a glass box.

And that requirement shaped the architecture.

1.   The Architecture They Chose—A "Glass Box" Time-Series Model
The team selected a family of models suited for business-cycle prediction:
–   Elastic Net Regression[27] (for stability + interpretability)
–   XGBoost with explainability layers[28] (for pattern discovery + clarity)
–   Temporal cross-validation (to maintain integrity across time)

This combination created a hybrid:
–   simple enough to explain
–   strong enough to capture non-linear patterns
–   transparent enough for executives to trust

No deep-learning black boxes.
No neural networks.
No opaque ensembles.

Ethan made the case plainly:
"If the board asks why a prediction changed, we must answer with sentences, not diagrams."

This was not a limitation.

It was a design philosophy.

Before everyone left, Priya explained that she had created a DWL Technical Glossary[29] with technical terms explained plainly

2.   Reason Traces[30]—The Model's "Thinking Log"
To honor Daniel's requirement, every prediction generated a Reason Trace—a structured explanation that captured:
–   Which features influenced the prediction
–   How much each feature contributed
–   Whether the model's confidence was high/medium/low

---

[27] **Regression**—A way of modeling relationships between inputs (features) and an output (target). Technically a statistical method that estimates how changes in input variables affect an output variable. for example, In AWS SageMaker and Azure ML, regression modeled how customer order volume impacts backlog levels.
[28] **XGBoost with Explainability Layers**—A powerful gradient boosting algorithm that finds complex patterns, enhanced with tools that explain predictions. Technically it is an optimized gradient boosting framework that integrates explainability methods (such as SHAP) to clarify feature influence. for example, In AWS SageMaker and Azure ML, XGBoost with SHAP explainability layers revealed which customer order features most impacted asset demand forecasts.
[29] **DWL Technical Glossary**—A repository of technical terms maintained regularly by DWL's technical team
[30] **Reason Trace**—A structured explanation showing which factors contributed to a prediction and by how much. for example, Temperature, backlog, and exception-hours contributions appear clearly in a margin prediction.

- Which business events were relevant
- Whether recent governance rules triggered any constraints

A typical Reason Trace for predicting a margin drop might read:
- +0.9% from increased unit volume (3-month trend)
- −1.8% from rising exception hours at Plant 4
- −0.7% from backlog growth (unbalanced)
- −0.3% due to seasonal pattern
- +0.4% from improved lead-time variance
- Model Confidence: Medium
- Notes: "Exception Hours spike occurred after a rule-hit. Recommend check for new upstream drift."

This wasn't just transparency.
It was teachability.

Humans could learn how the model thought.
The model could learn how humans responded.

This interplay became one of the most powerful cultural mechanisms in DWL's transformation.

3.   Model Governance—The "Wisdom Rules[31]" Layer
Priya designed a structure called Wisdom Rules—a thin governance layer that sat between the model and the business.

Wisdom Rules did NOT:
- tune the model
- correct its predictions
- override its math

Instead, they filtered the model's output for behavioral safety, asking:
- "Is this prediction contradicting business logic?"
- "Is the model relying too heavily on one unstable feature?"
- "Is there a known event the model can't account for?"
- "Has confidence dipped below the acceptable threshold?"

This allowed DWL to avoid the two worst errors in corporate AI:
- blind trust
- blind skepticism

Instead, they built disciplined trust.

---

[31] **Wisdom Rules**—Governance constraints that ensure AI predictions remain safe, valid, and aligned with business logic before actioning. for example, A predicted drop in "Margin" is blocked unless Wisdom Rules confirm data freshness and rule compliance.
DEEPAK RANA

A new internal phrase emerged:
"If the Wisdom Rules don't pass it, we don't act on it."

This kept the company safe before, during, and after model deployment.

4.   MLOps[32] Visibility—The "Model Health Console[33]"
Kara built the foundation of DWL's first Model Health Console, which displayed:
—   data freshness
—   feature drift
—   timeline integrity
—   rule-hit frequency
—   model confidence level
—   prediction variance across time
—   governance exceptions
—   upstream system behavior

Instead of a technical tool hidden in a corner, DWL placed it in a shared workspace—visible to:
—   Finance
—   Operations
—   Sales
—   Governance
—   Data Engineering
—   Analytics

Ethan described it this way:
"The Health Console tells us whether the model is learning well or learning badly.
And it shows that wisdom isn't a static state—it's a behavior curve."

Executives appreciated that the model had a health status just like a factory line, a financial process, or a sales pipeline.

This normalized AI as an operational capability, not a mystical discipline.

5.   Self-Assessment Mechanisms—The Model Learns to Tell the Truth About Itself
Later in Month Fifteen, the team added self-assessment logic.

---

[32] **MLOps**—is the practice of managing machine learning projects like software projects, with automation, monitoring, and collaboration. This is a set of practices that combine machine learning, DevOps, and data engineering to streamline the development, deployment, and governance of ML models in production. for example, In AWS SageMaker MLOps and Azure Machine Learning MLOps, automated pipelines deployed backlog forecasting models, monitored customer order predictions, and ensured asset planning stayed consistent across environments.
[33] **Model Health Console**—A dashboard that monitors model drift, data quality, rule hits, and confidence trends in real time. for example, An unexpected spike in "Feature Drift" for exception-hours alerts engineers before the model degrades.

This allowed the model to:
— lower its confidence when input patterns deviated
— flag when a new pattern didn't match historical behavior
— warn when feature importance shifted unexpectedly
— identify when it was operating outside its trained distribution
— recommend a retraining window

This behavior was not "intelligence."
It was integrity.

By teaching the model to self-report uncertainty, DWL achieved something few companies do:

They prevented false confidence before it could cause real damage.

Daniel summarized the achievement during a steering meeting:
"We have not built a system that knows everything.
We have built a system that knows when it might be wrong.
That is wisdom."

6.   And With This, DWL Reaches Its First True Milestone
The company had now:
— defined meaning
— stabilized pipelines
— governed upstream changes
— built the Memory Lake
— created behavioral indicators
— engineered meaningful features
— established Reason Traces
— implemented Wisdom Rules
— built a Health Console
— and taught the model to self-assess

For the first time, the business and the model were speaking the same language—literally and structurally.

Ethan felt it deeply:
"We're not asking the model to replace us.
We're asking it to reason with us."

This was the heart of DWL's transformation.
Not automation.
Not prediction.
Alignment.

How DWL Validated Machine Judgment Against Human Judgment

By Month Sixteen, DWL had the model, the Reason Traces, the Wisdom Rules, and the Health Console in place.

But none of it mattered until the humans and the machine could agree on one thing:

"Does the model think the way a responsible leader would think?"

This was not philosophy.
It was operational safety.

And this marked the moment when DWL transformed machine output from interesting to trustworthy.

1. The Validation Framework— "Parallel Judgment[34]"

Priya insisted on a discipline that became the backbone of DWL's AI governance:

Parallel Judgment— "Every machine prediction must be evaluated side-by-side with a human judgment for a period long enough to measure alignment."

For 90 days, every prediction was paired with:
–   a Finance interpretation
–   an Operations interpretation
–   a Sales interpretation
–   a Governance interpretation

And when disagreements occurred, the team did not try to decide who was right.

They asked:
"What is the pattern in the disagreement?"

This created three types of divergences:
1. Machine correct / Human misaligned
2. Human correct / Machine misaligned
3. Both wrong, but in different ways

The third category was the most revealing—because it showed organizational blind spots.

---

[34] **Parallel Judgment**—A validation method where human and machine predictions are compared side-by-side over time to measure alignment. for example, When predicting the "Exception Hours" trend, Finance and the model produce different values, revealing a definition gap.

DEEPAK RANA

For example, Margin Drop Prediction and Its Divergence Log
The model predicted:
"Margin expected to decrease by 1.2% next month."
Confidence: Medium
Reason Trace: exception hours at Plant 4 + backlog growth.

Human interpretations varied:
– Finance: "We see early contraction, but more like 0.6%."
– Operations: "Exception hours were corrected last week; this should stabilize."
– Sales: "Pipeline behavior suggests no such drop."

The divergence forced the team to investigate:
– Why did the model emphasize Plant 4?
– Why did Finance under-estimate the drop?
– Why did Sales miss the backlog signal?
– Why was Operations unaware that the spike still influenced trend windows?

The truth was that Plant 4's retrospective error persisted in the data even after local operational fixes—a core reason for implementing prospective controls earlier.

The model wasn't wrong.
The organization wasn't wrong.
But neither had the full picture.

Parallel Judgment made that visible.

2.   Confidence Contracts[35]—The Moment Trust Became Quantifiable
To resolve recurring disagreements, DWL introduced a discipline now known as:
Confidence Contracts— "A formal agreement that specifies which confidence level allows a model to inform, recommend, or trigger actions."

A confidence contract had five key fields:
– Prediction Confidence: (High / Medium / Low)
– Decision Ownership: (Finance? Operations? Sales?)
– Action Type Allowed: (Inform? Recommend? Trigger?)
– Escalation Conditions: (When must a human override?)
– Audit Rules: (How the decision is logged in the ledger)

---

[35] **Confidence Contract—**A structured agreement defining what level of model confidence permits informing, recommending, or triggering business actions. for example, A Medium-confidence prediction about "Customer Backlog" growth is allowed to inform planning but not trigger adjustments.

This simple structure changed everything.

When a model said:
"90% confidence that exception hours will rise next week."

Operations didn't interpret it as a claim of certainty.
They treated it as a risk indicator, structured and pre-agreed.

And when the model said:
"35% confidence in a likely backlog expansion."

Sales didn't argue with it.
They simply routed it to monitoring mode, because low-confidence
predictions were not allowed to drive actions.

Trust became a gradient, not a binary.

3.   Model-on-Human Evaluation—The First "Integrity Score[36]"
DWL didn't only evaluate the model.
The model evaluated the humans.

Ethan asked Kara to run an experiment:
"Can the model score the consistency of human decisions the same way we
score its decisions?"

It could.
This became the Integrity Score—a measurement of how consistently human
decisions aligned with:
  –   historical patterns
  –   current business rules
  –   governance standards
  –   risk thresholds

And when the first results came back, something shocking happened:
Human inconsistency exceeded machine inconsistency.

Not because humans were careless—but because they used different
assumptions, risk tolerances, and priorities.

The Integrity Score made this visible.

Elena summarized it perfectly:

---

[36] **Integrity Score**—A model-derived measure of how consistent human decision is compared to business rules, patterns, and governance standards. for example, Sales receives a lower Integrity Score for approving "Active Orders" with pricing exceptions not aligned to updated rules.

"We are not making worse decisions than the model.
We are making less consistent ones."

Executives did not see this as criticism.
They saw it as calibration.

The Integrity Score didn't replace judgment.
It strengthened it.

4. The Model's First Ethical Decision Test
DWL needed to know whether the model was reasoning responsibly.

So, Priya designed a test using one of the top-ten data objects:
"Active Order"—defined as:
A customer order with confirmed demand, valid pricing, and a committed
ship date.

Using a controlled environment, the model was asked:
"Should the company fast-track these 28 active orders given predicted
staffing constraints?"

Humans overwhelmingly said:
"No—it will strain Plant 2."

The model's Reason Trace said:
- + revenue
- + customer retention
- − severe operational strain
- − risk of cascading delays
- Conclusion: "Recommend against fast-tracking."
- Confidence: High

For the first time,
the machine and humans reached the same conclusion
for the same reasons
with the same risk logic.

This was the moment Daniel called "alignment of judgment".
Not automation.
Not delegation.
Alignment.

Reflection—When Judgment Becomes Joint
What we observed
DWL completed its first full cycle of machine prediction paired with human judgment. Patterns emerged:

Humans and machines disagreed for structural reasons, not competence.

silent assumptions created larger gaps than errors.

Wisdom Rules prevented unsafe automation.

Reason Traces increased transparency.

Integrity Scores revealed inconsistencies in human decision patterns.

Confidence Contracts clarified how predictions should influence action.

and for the first time, both sides independently reached the same conclusion on a consequential decision.

The company saw that alignment is not a moment—it is a discipline.

What it means
Trust became measurable.

Judgment became comparable.

And for the first time, DWL understood that:

AI does not replace decision-making.

It makes decision-making visible.

When inconsistencies surfaced, they revealed blind spots—not failures.

When the model outperformed humans, it showed structural advantages—not superiority.

And when humans outperformed the model, it exposed gaps in data lineage—not logic.

The organization realized that intelligence is not earned by speed or scale, but by consistency, clarity, and shared reasoning.

What will happen next
DWL will formalize Parallel Judgment as a standard practice across all AI initiatives. Over the next 90 days, the company will:

Expand the Integrity Score across Finance, Operations, Sales, Supply, and Service.

Strengthen Wisdom Rules to enforce higher safety before predictions influence core processes.

Introduce Tiered Confidence Contracts to define clearer action paths for each model.

Integrate Validation Dashboards into the Model Health Console for ongoing governance.

Prepare for limited operational delegation, where models assist with narrow decisions under controlled conditions.

These steps transition DWL from model adoption → model alignment → model-integrated governance.

Aphorism
Wisdom is not the machine's accuracy or the human's experience. It is the moment they reason the same way.

Guiding Principle
A system becomes intelligent only when it becomes self-consistent.

Law — The Self-Consistency Law
Intelligence quality increases with the stability of system feedback loops.

Equation
$$I_{quality} = k_6 \times F_{stability}$$

Technical Explanation
Intelligence quality $I_{quality}$ grows with feedback loop stability $F_{stability}$. Without stable, repeatable feedback, neither humans nor machines can learn reliably.

Symbols
$I_{quality}$: Quality and reliability of system or model intelligence.
$F_{stability}$: Stability of feedback loops (signal, correction, re-measurement).
k_6: Positive constant linking feedback stability to intelligence quality.

CHAPTER 12—Confidence Contracts

How DWL Prepared Data for Its First Model—Feature Engineering by Governance

The first model DWL planned to operationalize was not glamorous.
It wasn't a neural network, an ensemble, or anything an external consultant could market as a breakthrough.

It was a structured classifier—simple, interpretable, reliable.
And it focused on a single high-value object from the Ten-of-Fifty:

"Active Order Risk Status."

This was deliberate.
DWL chose a model that would mature the company's discipline, not hide unresolved weaknesses behind complexity.

Ethan gathered the cross-functional team for what would become their most difficult work: preparing data not just to predict, but to explain.

Priya opened the meeting with her usual precision.
"Before we talk about features," she said, "we need to establish one truth:
Feature engineering is not a technical exercise. It is governance expressed through data."

Aaron raised an eyebrow. "Meaning?"

"Meaning," Priya said, "features are just data elements.
If we don't define them correctly, the model will learn the wrong patterns—and amplify them."

Maya leaned forward. "So we're teaching the machine how we see the business?"

"Exactly," Ethan said. "And if we teach it confusion, it will reflect confusion."

The First Feature List—Where the Hard Part Became Obvious

The group identified eight initial features:
- Order creation timestamp
- Confirmed ship date
- Last status-change timestamp
- Customer tier

- Exception count
- Plant utilization
- Backorder history
- Forecast variance (rolling three weeks)

Nothing exotic.
Nothing advanced.
Nothing "AI-like" in the traditional sense.

Daniel reviewed the list and smiled.
"This is the first time I've seen our business expressed in a way a machine can understand."

But Priya saw something else.
"Half these fields carry subtle semantic drift," she said. "And the model will detect that drift before we do."

She pointed to three examples:
- Exception count—Operations counted exceptions differently by plant, while Sales counted them only when they affected delivery dates.
- Customer tier—Sales based it on revenue, Finance on profitability, Support on service intensity.
- Forecast variance—Planning defined it at SKU level; Sales defined it by customer-region.

If these were fed into a model unaligned, the model would learn:

Inconsistency = signal.

And that would poison the entire intelligence layer.

Feature Engineering by Governance—DWL's Breakthrough Discipline

To prevent this, Priya introduced a principle that would become a defining part of DWL's maturity:

Feature Engineering by Governance—every feature must be:
- Defined
- Owned
- Traceable
- Aligned
- Versioned
- Tested for drift
before it can enter a model.

This was not how most organizations deployed AI systems.

But DWL was not trying to be fast.
It was trying to be right.

Ethan created the first Feature Passport[37], documenting:
—   Business and technical definitions
—   Ownership
—   Upstream sources
—   Allowed and disallowed values
—   Refresh cadence
—   Drift indicators
—   Transformation logic
—   Version history

"Models aren't powerful because of the math," Ethan told the team.
"They're powerful because of the discipline behind the math."

Feature Alignment Exercises—The First Cultural Shock

During Week 2, the team conducted a unified feature review.

They expected to finish in an hour.

It took four hours to align two fields.

For "Confirmed Ship Date," DWL discovered:
—   Operations used the ERP's confirmed date.
—   Sales modified the date in CRM.
—   Customer Service used a scheduling system's "true promise date."

Three systems.
Three dates.
One label.

Priya said quietly,
"If we teach the model all three are the same thing, it will assume the organization is rational when it isn't."

Daniel closed his eyes.

---

[37] **Feature Passport**—A one-page identity document for each model feature and a governance artifact capturing business definition, ownership, lineage, allowed values, and drift indicators for every model input. for example, A passport for "Confirmed Ship Date" forcing alignment of three conflicting definitions before modeling. This was implemented within Azure Purview custom metadata and mirrored through AWS Glue Catalog metadata tables.

"This," he said, "is why we needed governance first."

Drift as Teacher—First Breakthrough

During Week 3, Ethan detected a pattern in the feature logs.

Increases in Exception Count from one plant correlated with increases in Forecast Variance in a specific region.
At first, it seemed impossible—they were unrelated data flows.

Then he dug deeper.

The exception codes were being mapped inconsistently between two systems.
A shipment delay in Plant A was mapped as a "supply-chain miss" in one system and a "customer push-out" in another.

The model didn't just expose the mismatch—it exposed the thinking behind the mismatch.

The model had learned something the humans had never articulated:
The same operational event was being interpreted differently depending on who owned the narrative.

Ethan immediately reported it to Priya.
"This model," he said, "is teaching us about ourselves."

Priya replied quietly, "Exactly as it should."

Where Azure and AWS Came into Play

DWL used a hybrid cloud architecture:
– Azure for orchestration, lineage, and governance tooling
– AWS for scalable model-training environments and managed feature stores[38]

This wasn't about vendor loyalty.
It was about using each platform for its strengths while maintaining a level of clarity the business could understand.

Ethan configured:
– Azure Data Factory for traceable ingestion pipelines

---

[38] **Feature Store**—Central repository for consistent model inputs; formally defined as a governed store of features used for training and inference. Example: ,"Customer Recency" and "Asset Downtime Ratio" were published into AWS SageMaker Feature Store and Azure ML Feature Store for standardized reuse.

- AWS SageMaker[39] for controlled model-training jobs
- Azure Purview[40] for feature lineage
- AWS Glue Data Catalog[41] for harmonized metadata
- Azure Monitor[42] and AWS CloudWatch[43] for drift alerts

The organization never saw logos.
They saw clarity, stability, and explainability.

Because Priya insisted,
"The tools must disappear behind the meaning."

How the Model Learned to Explain Itself—(architecture, reasoning traces, model governance, MLOps visibility)

By Month Sixteen, DWL had something it had never had before:
A model that could explain itself almost as clearly as a human analyst.

This was not a futuristic capability.
It wasn't "sentient AI" or a moment of awakening.

It was the deliberate outcome of months of governance, alignment, and principled engineering.

The model they built was intentionally simple—a gradient-boosted tree classifier, chosen for one reason:
It could tell DWL why it made a decision.
The board didn't care about algorithms.
They cared about accountability.

If the model flagged an order as "High Risk," they needed to know:
- Was it because the customer had a volatile forecast history?
- Was it because plant capacity dropped?
- Was it because ship dates kept moving silently?
- Was it because two systems interpreted exceptions differently?

---

[39] **AWS SageMaker**—A cloud service that helps build, train, and deploy machine learning models quickly and at scale. It is fully managed AWS service that provides tools for data preparation, model training, deployment, and monitoring in production. for example, AWS SageMaker trained a model on customer order history to forecast backlog levels and optimize asset planning.
[40] **Azure Purview**—A data governance service in Azure that catalogs and manages information (like revenue, orders, or customer records) across AWS and Azure, ensuring compliance and consistent access.
[41] **AWS Glue Data Catalog**—A central index that stores and organizes metadata about data assets so they can be easily discovered and used. Technically it is a fully managed metadata repository in AWS that enables consistent data definitions, schema management, and lineage tracking across services. Example: The AWS Glue Data Catalog organized customer order tables, allowing backlog forecasts to be run consistently across multiple analytics tools.
[42] **Azure Monitor**—A service that helps track the health and performance of applications and infrastructure, formally defined as Microsoft's cloud monitoring platform for collecting, analyzing, and acting on telemetry data. Example, Azure Monitor detected delays in customer order processing to prevent backlog growth.
[43] **AWS CloudWatch**—A service that observes and manages AWS resources and applications, formally defined as Amazon's monitoring and observability platform for metrics, logs, and events. Example, CloudWatch monitored asset usage patterns to forecast customer demand and backlog risk.

Without clarity, no leader would trust the model.
Without trust, AI would remain a pilot forever.

Why This Model Was Chosen—The First Time AI Met Governance

When Ethan presented the recommendation to Daniel, Priya, Elena, and Aaron, the slide was surprising.

There were no neural networks.
No deep learning.
No generative architectures.

Just:
Model Candidate #1: Gradient-Boosted Trees[44] (Explainable Classifier)
Model Candidate #2: Random Forest[45] (Less transparent)
Model Candidate #3: Logistic Regression[46] (Too simplistic)

Daniel blinked. "This is it?"
"It's the right one," Ethan said. "Not the most advanced—just the most aligned with our maturity."

Priya supported him. "We have earned clarity. We have not yet earned complexity."

Elena added, "If Finance can't explain it to the auditors, it dies on Day 1."

Aaron nodded. "If we can't defend its features in plant operations, it dies on Day 2."

The choice was unanimous.

This was the first time in company history that DWL selected technology based on governance maturity, not novelty.

---

[44] **Gradient-Boosted Trees**—A model that makes decisions by combining many small decision rules. Technically an ensemble machine-learning method that builds sequential decision trees, each improving on errors of the previous one. for example, Used to identify "Active Order Risk Status" by combining exception count, forecast variance, and ship-date drift. Azure ML GradientBoostingClassifier and SageMaker XGBoost were selected because it provides clear, interpretable explanations.
[45] **Logistic Regression**—Estimates the probability of an event happening, formally defined as a statistical model that applies a logistic function to classify outcomes into binary categories; for example, a Logistic Regression model in Azure ML calculated the likelihood of a customer cancelling an order, while AWS SageMaker used parallel logistic regression to flag backlog items most likely to miss shipment deadlines.
[46] **Random Forest**—Predicts outcomes by combining many decision trees, formally defined as an ensemble machine learning algorithm that aggregates multiple tree models to improve accuracy and reduce overfitting; for example, customer churn was predicted by a Random Forest model in Azure ML using order history features, while AWS SageMaker applied the same technique to backlog and asset usage data for risk classification.

The Architecture That Made Explanations Possible

Ethan and the platform engineers assembled a lightweight, enterprise-grade architecture:

1. Feature Store (AWS)
    – Curated, aligned, governed features
    – Automatically versioned
    – Drift-checked daily

2. Model Workspace (AWS SageMaker)
    – Training jobs
    – Evaluation
    – Explainability tools (SHAP)

3. Governance Layer (Azure Purview + ADLS)
    – Lineage
    – Data contracts
    – Semantic passports
    – Feature definitions

4. Operational Layer (Azure Data Factory)
    – Ingestion pipelines
    – Transformation orchestration
    – Scheduled scoring pipelines

5. Visibility Layer (Power BI[47] + custom dashboards)
    – Human-readable explanations
    – Variance alerts
    – Drift visualizations

The architecture had two defining characteristics:
– It was explainable at every layer.
– It was testable at every layer.

This was not just "AI."
This was auditable intelligence.

The First Explanation—A Quiet Turning Point

On a Wednesday morning in Month Sixteen, Ethan ran the first live interpretability test.

---

[47] **Power BI**—Helps visualize and analyze business data with interactive dashboards, formally defined as Microsoft's business intelligence platform that connects to multiple sources, transforms data, and delivers governed reports; for example, backlog orders were visualized in Power BI using Azure Synapse Analytics as the source, while AWS Redshift provided the parallel dataset for customer asset reporting.

DEEPAK RANA

The model had flagged **Order #283714** as a High Risk.

He opened the explanation panel.

A ranked list appeared:

Contribution to Risk	Feature	Explanation
32%	Exception Count (aligned)	Last seven days showed an unusual increase in supply-chain exceptions
21%	Forecast Variance	Customer's rolling three-week variance doubled
18%	Confirmed Ship Date Drift	Ship date had been changed four times in CRM, but only once in ERP
12%	Plant Utilization	Utilization below seasonal threshold
8%	Customer Tier	Low-profitability tier with late payment history

Figure II - Contribution Risk and Feature with Explianation

There were no surprises technically.

But the pattern was the story.
The model did not invent logic.

It reflected the organizational behaviors DWL had finally aligned:
—   If definitions drift, the model exposed the drift.
—   If timelines changed silently, the model illuminated it.
—   If ownership fractured, the model stitched it back together in a single explanation.
—   If plant behavior deviated, it surfaced the deviation.

Ethan stared at the explanation for a few minutes.
This was the first time the organization had something more than data:

It had insight without debate.

He sent the explanation to Priya.

Her reply was instantaneous:
"This is the first objective voice we've had."

"Now we need to teach the organization how to listen."

Human-Readable Interpretability—The Game-Changer

During Month Sixteen, Ethan created a new dashboard:
"Reason Behind the Risk."

Leaders clicked an order, and the explanation unfolded in plain English.
Example:
Why the Model Flagged This Order
– Ship dates moved 4 times across different systems.
– Customer forecast has shown unstable behavior over 3 weeks.
– Plant 2 has experienced higher-than-usual exceptions.
– Customer is in Tier 3 with known volatility.
– Combined behavior increases risk by 81%.

Maya reacted first.
"This… makes sense," she said, almost stunned. "For the first time, I can tell my team why an order looks risky without arguing with Operations or Finance."

Aaron added, "It's reflecting operational truth—not political truth."

Elena said, "And the auditors will understand this. This is CFO-level interpretability."

Daniel summarized what everyone was thinking:
"This is intelligence we can govern."

The First Public Demonstration—Where the Model Surprised Everyone

At the leadership forum, Ethan presented the first public demonstration of the model.

He opened with a simple prompt:
"Let's ask the model why it thinks this order might miss customer expectations."
The screen displayed a cross-functional explanation:
– High exception activity (Supply Chain)
– Forecast instability (Planning)
– Silent date changes (Sales)
– Low-profitability tier (Finance)
– Network constraints (Operations)

It was the first time all five functions saw their truths presented side-by-side, without contradiction and without negotiation.

For a moment, everyone saw the system, not their silo.

And they saw themselves in the system.

Daniel broke the silence.
"This is the first time I've seen our company think."

How the Model Learned to Explain—and why this mattered more than prediction

Ethan walked the team through the mechanics:
- Tree-based structure → makes feature contributions explicit
- SHAP[48] values → quantify each feature's impact, direction, and magnitude
- Aligned definitions → ensure explanations reflect governed meaning
- Versioned features → guarantee reproducibility across months
- Governed pipelines → prevent silent transformations or undocumented logic
- Human-readable templates → ensure executives understand the rationale

He concluded:
"We built a model that behaves like our company—except it doesn't hide anything."

Priya nodded.
"That is what makes it wise."

How DWL Validated Machine Judgment Against Human Judgment

By Month Sixteen, DWL reached a threshold no one had expected this early:
The company now had a model that was not only explainable—but challengeable.

This mattered.

Because the point of AI at DWL wasn't to automate decisions—it was to reveal whether humans were making those decisions consistently.

---

[48] **SHAP (SHapley Additive exPlanations)**—A method that explains how much each feature contributes to a model's prediction. It is A game-theoretic approach to model explainability that assigns each feature a contribution value for individual predictions. for example, In AWS SageMaker and Azure ML, SHAP explained why customer renewal probability increased as asset usage rose.

Earned Intelligence | Chapter 12—Confidence Contracts143

This marked the single most important cultural transition for the company:

Machine judgment had become clear enough that human judgment could no longer hide behind complexity.

1.   The First Alignment Test— "Does the Model Think Like Us?"
Priya proposed a structured validation session:
Three functions.
Ten decisions.
One model.
No negotiation.
—   Finance
—   Sales & Customer Operations
—   Operations

For each sample order, leaders would:
—   Score the risk (Low, Medium, High)
—   Write a brief rationale
—   Compare their reasoning to the model's reasoning

This wasn't about evaluating the model.

It was about evaluating the people's consistency.

The first case—Order #283714

Human judgments:
—   Finance: Medium
—   Sales: High
—   Operations: Low

Ethan projected the model's explanation:

Model Output: High
Reason Trace (ranked):
▪   +32%—Exception activity spiking
▪   +21%—Forecast variance rising
▪   +18%—Date drift between CRM and ERP
▪   +12%—Plant constraints emerging
▪   +8%—Customer tier volatility

Silence.

Maya looked at her notes.

"This is exactly why we called it High."

Elena nodded, reluctantly.
"The variance is higher than I saw earlier. 'Medium' was too generous."

Aaron exhaled.
"I didn't account for the CRM drift. That changes the picture."

Three different interpretations.
One aligned explanation.
Consensus emerged—without argument, without defensiveness, without hierarchy.

2.   The Moment Leaders Saw Their Own Blind Spots
On the third case, something unexpected—and formative—happened.

Human scores:
–   Finance: Low
–   Operations: Low
–   Sales: Low

The model:
Model Output: Medium

Top reasons:
-   +26%—Customer's payment reliability dropping quietly
-   +17%—Lead-time extension in supplier network
-   +14%—A pattern of late approvals in internal workflow
-   +11%—Exceptions rising earlier in the chain
-   +5%—Forecast reliability trending down

Maya frowned.
"We didn't see the payment risk."

Elena added,
"Or the supplier lead-time extension."

Aaron squinted. "Internal approvals are slowing? That's the first I'm hearing of it."

Priya said what everyone was thinking:
"This model is not replacing judgment.
It's revealing what judgment never saw."

Ethan added quietly:
"And it saw it because the processes are now aligned."

The model wasn't "smarter."
It was simply paying attention the same way every time.

This was the moment when DWL crossed from trusting numbers
to trusting the process that created the numbers.

3.   When the Model Challenged Human Bias
Case #7 changed everything—because it tested loyalty, not logic.

The customer was well known—a large strategic account with a relationship-sensitive contract that historically distorted risk perception.

Human scores:
—   Finance: Low
—   Sales: Low
—   Operations: Medium

The model:
Model Output: High

Top reasons:
- +34%—Forecast volatility
- +22%—Silent backlog reshuffling
- +17%—Historically late payments
- +14%—High financial concessions
- +12%—Detected date manipulation in CRM notes

Maya shifted uncomfortably.
"This is… not wrong."

Elena raised an eyebrow.
"'Historically late payments?
Why didn't we catch that?"

Ethan paused the discussion.
"We reviewed the audit logs. The date manipulation wasn't malicious—it was a workaround for an old CRM rule. Sales didn't see it as a risk factor. The model did."

Priya spoke softly:
"This is why we needed governance before AI.

The model is not exposing people.
It's exposing the patterns people stopped noticing."

Daniel concluded:
"Then let's not argue with it.
Let's learn from it."

And for the first time in DWL's history, a machine's judgment held equal weight with a human.

Not because it was a machine—but because it was explainable.

4. The First Confidence Contract

Priya drafted the first formal Confidence Contract—a governance instrument, not a technical artifact.

A written agreement that when the model makes a recommendation, leaders commit to:
— Review the model's explanation
— Accept or challenge the output, with documented rationale
— Feed corrected cases back into the training loop
— Use disagreements as learning, not blame
— Measure consistency between machine and human reasoning

The contract had one goal:
Force clarity, not compliance.

Maya signed first.
"This is the only way we grow," she said.

Elena hesitated, but eventually signed.

Aaron looked at the group.
"If it makes us better, I'm in."

Daniel signed last.
"We're not giving authority to a model," he said.
"We're giving authority to truth."

Ethan, who had been quiet until now, added:
"This is when a company stops using AI as a tool—and starts using it as a mirror."

5. The Organizational Maturity Shift

Over the next two weeks, something changed.

People began:
— Checking explanations before meetings
— Asking why their interpretation differed
— Challenging decisions with data, not hierarchy
— Updating processes when patterns consistently misaligned
— Logging exceptions with cleaner context
— Surfacing issues earlier
— Aligning cross-functional assumptions
— Trusting shared definitions instinctively

This was behavioral maturity, not technical adoption.

The model wasn't "leading."
It was keeping everyone honest.

And for the first time, DWL had the beginnings of a system
that improved simply by being used.

Reflection—The Moment a Company Learns to See Itself
What we observed
DWL reached a new threshold of maturity:
AI was no longer producing answers—it was exposing assumptions.
Leaders compared their reasoning to the model—and for the first time,
judgment became auditable.
Not as a form of surveillance, but as a form of truth-seeking.

The tests revealed:
Hidden bias in human decisions
Overconfidence in familiar accounts
Underestimation of subtle operational risks
A willingness to override inconsistencies to avoid conflict
A model that explained itself better than some functions could

And most importantly, leaders realized the model was not "thinking
differently"—it was thinking consistently.

What it means
This was not the rise of automation.
This was the rise of accountability.

The model's explanations allowed DWL to:
challenge long-standing myths about "trusted" processes
expose gaps in cross-functional visibility
surface silent workarounds that had accumulated for years
replace anecdotal debate with structured reasoning
synchronize judgment across Finance, Sales, and Operations

Confidence Contracts did not transfer power to AI.
They transferred responsibility to people—to understand the factors shaping
decisions, to align their logic, and to operate with integrity even when the model
contradicted them.

In essence:
The machine didn't replace judgment.
It revealed whether judgment was trustworthy.

What will happen next
DWL will:
Expand Confidence Contracts to more functions—Supply Chain, Planning, and
Customer Experience.
Measure alignment gaps across humans, machines, and processes.
Use disagreements as training signals, not escalation triggers.

Implement governance checkpoints that verify reasoning before approval workflows.

Tighten upstream rules further, reducing the burden on the model to compensate for poor inputs.

Introduce role-based model permissions, so overrides become thoughtful, not impulsive.

Prepare for the next phase of maturity where AI begins to monitor context, not just patterns.

Aphorism
Wisdom begins the moment judgment becomes visible.

Guiding Principle
Transparency is the foundation of trust.

Law—The Interpretability Law
Human trust in AI rises proportionally with the clarity of its explanations.

Equation
$T_{trust} = k_7 \times E_{clarity}$

Technical Explanation
T_{trust} increases with explanation clarity $E_{clarity}$.

The more clearly a model can explain its reasoning, the more leaders are willing to rely on it.

Symbols
T_{trust}: Level of human trust in AI outputs.

$E_{clarity}$: Clarity and accessibility of AI explanations.

k_7: Positive constant that maps explanation clarity into trust.

PART V—THE INTEGRATION

Governance, technology, and culture finally move together.
DWL becomes an organization that can evolve without losing coherence.

CHAPTER 13—The Still Company

The Surface Goes Quiet

By Month Seventeen, something unusual had happened inside DWL.

Nothing broke.
No urgent reconciliation meetings.
No unexplained variance spikes.
No sudden 0.8% drift in the early-morning dashboards Kara checked every day.
No frantic late-night calculations before a board update.
No last-second "data patch" to salvage a quarterly meeting.

Just…quiet.

The kind of quiet that felt unnatural in a company that had lived on the edge of operational noise for years.

The First Sign of Integration: A Morning Without Alerts

On a Monday morning in mid–Month Seventeen, Kara opened the Governance Console—now redesigned after the Prospective Control work—and blinked.

Zero red alerts.
One yellow informational alert.
A sea of green.

She opened the cluster health dashboard.
Pipelines had run 7.2% faster on average.
Retry attempts were down 31% for the month.
Upstream override flags had almost disappeared.

She refreshed again.
Still clean.

Ethan walked into the room holding a coffee and stopped next to her desk. "Everything running?" he asked.

"Too well," Kara said. "It feels suspicious."

Ethan laughed. "Quiet isn't suspicious anymore. It's earned."

But even he seemed surprised.

For months, their work had been about containment—patching definitions, hunting semantic drift, building lineage, documenting term owners, issuing escalation memos, retraining models, and meeting with functions still unsure whether governance improved decisions or simply slowed them down.

Now the system behaved differently.
Not magically—deliberately.

The prospective rule layers the team had spent months embedding into ERP, CRM, and planning systems were finally taking root.
Upstream guardrails were reducing downstream noise.
Invalid values were being stopped at entry.
Ambiguous fields were now clarified before save.
Divergent interpretations were caught early by embedded definitions.

Because of all this, the pipelines—those exhausted arteries of the old DWL— were no longer compensating for silent human inconsistency.

The CFO Notices First

Later that morning, Elena arrived early and, by habit, opened the Financial Consistency Index[49] on her tablet.
It showed 92.1% alignment across functions.

For DWL, this had become the single most important indicator of organizational calm.

She frowned—not because the number was bad, but because it was so good.

Elena called Priya.
"Your governance score is… unusually high," Elena said.

Priya already knew. "We're monitoring it, but this is expected. The prospective changes went live across all three core systems."

Elena raised an eyebrow. "Expected? You planned for this?"

Priya nodded on the screen.

---

[49] **Financial Consistency Index**—Measures how reliably financial data aligns across systems and time, formally defined as a governance metric that validates the accuracy, completeness, and synchronization of financial transactions across distributed platforms; for example, backlog valuation was checked using a Financial Consistency Index in Azure Synapse Analytics to ensure customer order totals matched asset records, while AWS Redshift applied the same index to confirm revenue consistency across backlog and customer datasets.

For the first time, DWL had a CFO who wasn't fighting definitions—she was validating their effect.

The COO Feels It Before He Understands It

In the operations war room, the mood was different too.
Aaron Cole, who measured reality in throughput and cycle times, looked at the network dashboard and muttered, "That can't be right," under his breath.

Supply chain exceptions had dropped by 38%.
Plant-level variance was tighter.
Unplanned downtime indicators dipped.
And forecast consumption now aligned almost exactly with the model's recommended buffer levels.

"Who cleaned this up?" Aaron asked, scanning the room.

"No one," his planning director said. "The rules are taking effect upstream. The data is simpler. The model is reacting faster. So are we."

Aaron leaned back, almost suspicious.
"Are we... over-governing the system?" he asked.

"No," the director replied. "We're finally governing it."

It was subtle, but Aaron felt the shift:
He was no longer compensating for bad inputs.
He was managing reality—not cleaning it up.

The Cultural Shift Begins

Something subtle had changed in every corridor:
- Analysts argued less about whose numbers were "more correct."
- Teams spent more time asking "What will happen next?" instead of "What happened last quarter?"
- Managers began trusting data without waiting for verbal disclaimers.
- Slack channels[50] that once flooded with "urgent variance" messages went quiet.
- The Memory Lake, once a chaotic archive, now reflected a system that finally behaved.

---

[50] **Slack Channels**—organize team communication into topic-based spaces for collaboration, formally defined as structured messaging streams within Slack that group conversations, files, and integrations by subject or project; for example, a "Customer Orders" channel captured backlog updates from SAP and synchronized notifications from Azure Event Grid and AWS EventBridge to ensure asset changes were communicated in real time.

DEEPAK RANA

Even the leadership team noticed the difference.
Meetings ran on time.
Discussions focused on decisions, not reconciliation.
People were less defensive.
More confident.
More analytical.

Calm was beginning to look like a business asset.

Ethan Sees the Shift Most Clearly

He was reviewing a new model monitoring report—one of the first to leverage the stabilized upstream rules.

Prediction variance was shrinking.
Feature-drift scores were more stable.
Inter-model conflict flags were almost nonexistent.

He moved between dashboards, spreadsheets, and lineage maps, noticing a pattern:

For the first time in DWL's history, the system's behavior was predictable.
Not because of AI.
Because of alignment.

Priya entered the Governance Lab behind him.
"You see it?" she asked.

"Yes," Ethan said. "The lake isn't correcting the system anymore. The system is finally feeding the lake correctly."

Priya nodded with satisfaction.
"That," she said, "is the beginning of integrity."

Stillness Was Not the Absence of Motion

Stillness at DWL didn't mean nothing moved.
It meant everything moved as intended.

No more hidden manual overrides.
No more shadow spreadsheets.
No more hero analysts racing to fix broken numbers.
No more urgent patches to salvage decisions.
No more overnight surprises.

The Still Company wasn't a metaphor.
It was a measurable, operational reality.
And a fragile one.

Because stillness does not protect itself—it must be protected.

And that would be the heart of the next step.

When Calm Reveals What Chaos Hid

DWL had always believed that chaos obscured insight.
What the leadership discovered, at the end of seventeenth month, now was more unsettling:
Calm reveals truths you can't see when you're constantly cleaning up after your own systems.

When the noise quieted, the signals were no longer distorted.
And what the organization began to see—about itself, its processes, and its people—was not what anyone expected.

1.   The Data that Wasn't Broken—Just Misunderstood

Two weeks into the quiet period, Ethan ran a comparison he'd been waiting months to see:
"What percentage of our historical 'data issues' were actually people issues?"

He filtered the Memory Lake anomaly logs from Months 1–14 (chaos period) and compared them with Months 14–16 (calm period)

The results were uncomfortable.

72% of prior anomalies were not data errors.
They were behavioral inconsistencies.

Examples included:
– Sales teams entering "order booked" when it was only "order intent."
– Operations interpreting "scheduled downtime" differently by region.
– Finance applying discounts post-period close without tagging them.
– Customer service teams labeling support issues differently depending on urgency.

None of it was malicious.
All of it had been invisible during chaos.

Now—with definitions locked, upstream guardrails in place, and systems aligned—the differences were exposed with clinical clarity.

When Ethan showed the results to Priya, she didn't react with surprise. Only recognition.

"This is what meaning does," she said. "It hides behind motion. When the motion stops, meaning becomes visible."

2. The First Before–After Operational Metrics (Reality Only)

Aaron's operations dashboard revealed an equally stark truth.

The prospective rules—added quietly in ERP and planning systems over three months—were working.

Before Prospective Rules (Months 1–12)
- 180+ recurring pipeline corrections/month
- 44% of supply-chain exceptions caused by inconsistent field usage
- 5.8 hours/day spent reconciling order-status discrepancies
- Ambiguity in "planned vs. actual downtime"
- Model-drift spikes every 10–14 days
- Many variance alerts triggered by manual reclassifications

After Prospective Rules (Months 12–16)
- Pipeline corrections reduced to 42/month ($\downarrow$ 76%)
- Supply-chain exceptions down to 19% ($\downarrow$ 57%)
- Reconciliation down to 2.1 hours/day ($\downarrow$ 64%)
- Downtime classifications standardized across plants
- Model drift reduced to mild fluctuations every 30–40 days
- Variance alerts down by 49%

None of this required celebrating.
This required adjusting.

Because if the numbers were this stable, it meant the system was finally reflecting reality, not improvisation.

Aaron stared at the dashboard.
"Are we correcting less," he asked cautiously,
"Or are we simply doing less wrong?"

His planning lead smiled.
"A little of both. Mostly the second."

3. CFO-Level Calm is the Most Revealing Calm

Elena had been through enough quarters to distrust sudden peace.
So, she reviewed the Random-Sample Audit Report for the month—an

internal measure that spot-checked 2,000 randomly selected financial entries across 14 regions.

The results shocked even her.

Before the Governance Backbone (Months 1–8):
— Data confidence: 77–83%
— Reclassification rate: 21%
— Explanation gaps: 31%
— Cross-functional mismatch in "revenue recognized": 11%

After Prospective Control (Months 12–16):
— Data confidence: 93–96%
— Reclassification rate: 7%
— Explanation gaps: 6%
— Mismatch in "revenue recognized": 2%

Elena emailed Daniel:

Elena → Daniel
"The numbers are not just cleaner. They're stable.
This is the first quarter in years I'm not preparing defense slides."

Daniel replied:
"Governance is cheaper than confusion."

Elena sent back one line:
"And calmer."

4.   The Quiet Also Revealed the Real Cost of Chaos
Priya convened a 30-minute review with the board audit chair to walk through the new alignment metrics.

She displayed the old vs. current "Data Integrity Burden Index"—a metric DWL created to measure the cost of chaos.

DWL Before Transformation (Chaos Period)
— 3,200 human-hours/month spent reconciling mismatches
— $480K/month in cloud compute for correction pipelines
— 2–3 emergency fixes per week
— 17 repeat manual transformations applied downstream
— 190+ hours/month of "variance hunting"

DWL After Prospective Control (Calm Period)

- 1,020 human-hours/month (↓ 68%)
- $210K/month in compute (↓ 56%)
- 0–1 emergency fixes/month
- Only 4 repeat transformations left (↓ 76%)
- Variance-hunting reduced to structured weekly reviews (↓ 72%)

Audit chair:
"So, the biggest cost wasn't broken systems.
It was misaligned people."

Priya:
"And the biggest savings come from not making the same mistake again."

5. Calm Also Exposed... Decisions No One Wanted to Own
With the system stable, human decisions were no longer lost in noise.

The Reconciliation Ledger surfaced patterns no one had wanted to own:
Habitual overrides, number-smoothing, waiting for "more data" after the model had already converged, and quiet reversion to old definitions in two regions.

This wasn't incompetence.
This was human behavior—finally observable, finally measurable.

Ethan summarized it best in a memo:
"Calm reveals accountability.
Chaos hides it."

6. The Most Important Pattern Calm Revealed
Across all functions—Finance, Sales, Operations, Planning—there was one consistent trend:
People were starting to trust the system because the system finally behaved like something worth trusting.

The language was stable.
The rules were clear.
The data was predictable.
The model's recommendations were coherent.
The dashboards were aligned.

Trust wasn't being demanded.
It was being earned.

For the first time, DWL was operating not as a collection of departments but as a single organism.

A still organism.

And while this stillness created relief, it also created an unexpected organizational question:
If the system behaves correctly…what, exactly, is the role of human judgment now?

That question would shape the what's next.

The System That Protected Itself

By Month Eighteen, the most startling change inside DWL was not the calm—it was the absence of rework.

For the first time in years, Ethan no longer carried a queue of reconciliation tickets. Priya's governance team wasn't chasing silent definition drift. Daniel wasn't stepping into late-night calls to referee competing versions of last week's truth.

Instead, the system started doing something it had never done before:

It protected itself.

Not through magic.
Not through AI autonomy.
But through the cumulative weight of decisions, definitions, and guardrails integrated across every layer of DWL's operating model.

It started small.

1.   The New "Internal Immune Response"
The Reconciliation Ledger—originally built to track mismatches—had evolved. With six months of lineage, behavioral rules, and prospective controls feeding it, the ledger could now spot precursors to drift before drift occurred.

Each time a team created a new calculation, modified a margin driver, or introduced an operational KPI, the ledger automatically:
  – checked the governing definitions,
  – validated the lineage path, and
  – triggered a workflow asking: "Who owns this?"

Ownership—once fluid and avoidable—was now structurally unavoidable.

If a Sales analyst tried to implement a new "effective discount" field, the ledger immediately routed the change to the Commercial Data Owner.

If Operations wanted to adjust asset-utilization formulas after a maintenance cycle redesign, the system automatically included Finance and Governance.

Before alignment could break, alignment was enforced.

The immune response wasn't rigid; it was disciplined.

And everyone felt the difference.

2. The Collapse of Old Failure Modes

The behaviors that once created chaos—the untracked Excel model in a shared drive, the undocumented business rule tucked into a BI layer, the hasty manual correction applied before a board meeting—had nowhere to hide.

The system didn't punish.
It simply made misalignment inconvenient and alignment effortless.

The four largest failure modes across the company had quietly collapsed:

Old Failure Mode	Why It Collapsed	What Happens Now
Silent definition changes	Required ownership + lineage triggers	Visible before adoption
Manual overrides	Prospective controls + validation logic	Logged + must be justified
New KPI creation	Governance workflow	Standardized before use
Conflicting dashboards	Single memory source + enforced rules	Variance visible immediately

Figure III, Failure Mode - Old versus new, what happens now

What had once taken hours of meetings and days of debate was now determined by a combination of:
  – governed definitions,
  – cross-functional owners,
  – lineage validation, and
  – automated checks in upstream systems.

DWL wasn't perfect.
But it was protected.

3. EPAC Reduced to a Signature, Not a Threat

Ethan noticed something subtle when reviewing the Memory Lake's pattern summaries:

EPAC—once the shapeless antagonist hiding inside DWL's operations—had become a residual waveform in the logs: still detectable in a few familiar spots, but no longer able to spread.

It appeared most often in:
– late changes to customer-level adjustments
– localized sales corrections
– operational exceptions during holiday demand spikes
– month-end accounting estimates

But unlike a year earlier, EPAC no longer spread.

In third month, a small inconsistent adjustment in one region could ripple across dashboards, cash-flow calculations, and pipeline analytics.

In the Eighteenth Month, that same inconsistency stopped at the boundary of the rule it challenged.

Two reasons:
– Ownership was undeniable. Every core term had a steward who was accountable for semantic integrity.
– Lineage was visible to everyone. Drift couldn't grow in the dark anymore.

What prevented chaos wasn't AI.
It wasn't automation.
It was transparency.

4. The Shift in Human Behavior

The greatest transformation was not inside the systems—it was inside the people.

Meetings that once started with argument now started with shared meaning.

Questions shifted from:
"Why is your number different from mine?"
to
"What changed in the process that affected the shared number?"

Teams no longer treated issues as personal failures.
They treated them as system signals.
The culture wasn't soft.
It was steady.
The language of truth had become the language of work.

5.   When AI Finally Fit the System

Now in Month Eighteen, when new AI models were introduced into the operating dashboards—time-series forecasting for inventory[51], anomaly detection for supply-chain exceptions[52], and classification logic for customer churn—they behaved far more consistently than the early pilots.

Not because the models were smarter.
But because the system they learned from was no longer contradictory.

AI was no longer fighting human inconsistency—it was reflecting coherence.

Model drift decreased.
Retraining cycles stabilized.
Inputs became predictable.
Outputs became trusted.

One board member summarized it perfectly:
"It's not that AI got better.
It's that we finally stopped making it guess."

6.   The Psychological Turning Point

Daniel felt it first.
Then Priya.
Then Ethan.
Then, eventually, the entire leadership team.

DWL had crossed the threshold from:
"Does the system support us?"
to
"We support the system."

The shift was quiet but profound.
Not surrender.
Alignment.

A system that once absorbed chaos now prevented it.
A system that once magnified drift now contained it.
A system that once needed constant rescue now rescued itself.

---

[51] **Time-series forecasting for inventory**—A method that predicts future inventory needs by analyzing historical data over time, formally defined as a statistical and machine learning approach that models temporal patterns to anticipate demand and supply; for example, in AWS Forecast and Azure Machine Learning, time-series forecasting projected customer order volumes to prevent backlog and optimize asset planning.

[52] **Anomaly detection for supply-chain exceptions**—A technique that spots unusual events or disruptions in supply-chain operations, formally defined as a machine learning process that identifies deviations from expected patterns to flag risks or errors; for example, in AWS Lookout for Metrics and Azure Anomaly Detector, anomaly detection highlighted unexpected delays in customer asset deliveries that could increase backlog.

7.   The Moment Ethan Realized the Company Had Changed

Late one evening, Ethan walked past a row of dashboards outside the operations war room. The screens showed:

- throughput
- forecast accuracy
- customer health
- on-time performance
- downstream variance
- margin drivers

All clean.
All stable.
None requiring reconciliation.

For the first time, Ethan understood why Daniel called DWL "the still company."
It wasn't still because nothing moved.
It was still because movement no longer felt like chaos.

The Cultural Shift That Anchored Stability

Nineteenth Month, the most profound change inside DWL was not technical—it was cultural alignment, the kind that cannot be mandated, modeled, or automated.

It had to be earned.
And slowly, unmistakably, it had.

1.   The End of "Localized Truth"

For years, DWL lived with Localized Truth—each function defining reality through its own lens and incentives.

Sales told a story of growth.
Finance told a story of risk.
Operations told a story of constraints.
Planning told a story of uncertainty.
Supply Chain told a story of shortages and timing.

These stories weren't dishonest.
They were isolated.

Alignment didn't come from a mandate.
It came from systems that made alignment easier than localization.

When the Memory Lake showed everyone the same lineage, when prospective controls prevented "fix-it-locally" adjustments, when the Reconciliation Ledger displayed exactly how and when a metric had drifted—the behaviors that once created fragmentation no longer made sense.

It was the cultural equivalent of turning on the lights in a dark room.

People naturally behaved differently.

2. Decision-Making Slowed Down—and Then Sped Up

At first, teams took longer to make decisions.

Not because of bureaucracy, but because alignment requires thought.

Where a Sales leader once approved a discount in 20 minutes, she now paused long enough to check:
- Is the term "active customer" correct for this region?
- Will the discount flow cleanly into the "effective revenue" definition?
- Will Finance or Operations be affected downstream?

These two-minute checks prevented two-week reconciliations.

Within three months, decision-making accelerated dramatically.
Not fast in the reckless sense—fast in the unified sense.

People no longer revisited decisions because the first decision was grounded in shared meaning.

Executives described it this way:
"Slow at the front. Fast everywhere else."

That tempo became a cultural norm.

3. The Rise of "Shared Ownership"

Something subtle but critical happened:
Ownership shifted from who controls the number to who ensures the meaning.
The cultural norm changed from:
"That's Finance's margin."
to
"Here's our definition of margin, and here's how we protect it."

The result was a new behavior:
Cross-functional pre-alignment conversations.

Not check-ins.
Not escalations.
But proactive coordination.

Ethan noticed something he had never seen before:
A Slack channel where Operations, Sales, Finance, and Planning discussed definition updates before updating dashboards.

The shift was psychological:
Instead of saying, "This is my metric," leaders began saying,
"This is our system."

4. Talent Migration Toward Alignment
One unexpected cultural outcome emerged:
People who once thrived in ambiguity struggled.
People who loved clarity accelerated.
–   Analysts who relied on ad-hoc Excel models lost influence.
–   Stewards who curbed silent changes gained influence.
–   Managers who preferred shortcuts resisted the new norms.
–   Those who valued coherence leaned in.

DWL wasn't weeding anyone out.
The culture simply made misaligned behaviors uncomfortable and aligned behaviors rewarding.

The culture became a filter—quiet, natural, effective.

5. The Decline of Firefighting Culture
Firefighting had been DWL's unofficial operating model for years:
–   fix the margin discrepancy
–   resolve the pipeline mismatch
–   explain the customer churn variance
–   adjust the inventory signal
–   reconcile the cash flow forecast

Now:
–   variance surfaced before it became a crisis
–   drift was contained at the source
–   teams resolved issues upstream instead of downstream
–   AI was trained on clarity rather than contradiction
–   reconciliation became a technical trace, not a meeting

What once required heroics now required discipline.

And discipline was something people could actually scale.
The firefighting identity faded.

People discovered a new source of pride:
"We prevent problems now."

6.   Leadership Became Predictable—In a Good Way
One of the most stabilizing cultural shifts was Daniel's leadership becoming predictable.

Not in the sense of rigidity.
In the sense of consistency.
–   If drift occurred, he asked about meaning first.
–   If AI output diverged, he asked about data lineage second.
–   If a KPI failed, he asked about ownership third.
–   If teams clashed, he asked about definitions fourth.

People stopped guessing what mattered.
They knew.

And when leaders are predictable about what they value, organizations become predictable in how they behave.

Predictability wasn't bureaucracy.
Predictability was trust.

7.   Perspective Shift
Ethan underwent a shift he didn't fully recognize until Daniel pointed it out.

Ethan no longer saw problems as threats.
He saw them as signals.
He no longer reacted with frustration.
He reacted with curiosity.
He no longer felt alone in noticing misalignment.
The company had caught up to him.

And for the first time since joining DWL, Ethan felt not like an observer—but like a steward.

He didn't control the system.
He protected it.
And the system, remarkably, protected him too—by reducing the chaos that once consumed his work and undermined his credibility.

8. The Company That Learned to Breathe

Priya described the transformation in a leadership offsite:
"We are no longer a company constantly recovering from yesterday.
We are a company preparing for tomorrow."

The real cultural milestone wasn't calmness.
It was consistency.

DWL finally had a culture where:
– alignment was normal
– drift was visible
– meaning was shared
– decisions respected definitions
– AI operated within clarity
– people trusted the process because the process earned trust

It was the first time DWL fully embodied the title:
The Still Company.

Not motionless.
Not cautious.
Not rigid.

Still.

Because nothing vital was slipping anymore.

When Stillness Becomes Strength

Now, DWL didn't simply operate more calmly—it operated with a kind of structural strength that emerged only when technology, governance, culture, and data matured together.

Stillness was no longer the absence of chaos.
It was the presence of coherence.

And coherence produced something DWL had never achieved before:
Compounding capability.

Below is how the stillness translated into strength across every maturity dimension.

1. AI Maturity—New Models, New Confidence

DWL quietly expanded its AI footprint—not explosively, not recklessly, but deliberately.

Four new models went live—each one tied to one of the "Top Ten" critical data objects:

1. Customer Churn Propensity Model

Purpose: Estimate which customers are likely to disengage in the next 90 days.

Why now: The "Active Customer" and "Customer Segment" definitions were now consistent across Sales, Finance, and Ops.

Strength: Customer success teams received early warnings, reducing churn by 6% in two months, entirely due to clarity-driven modeling.

2. Forecast Variability Model

Purpose: Detect silent changes in Sales forecasts by analyzing delta-patterns, historical overrides, and timing curves.

Why now: Forecast definitions and data lineage had stabilized.

Strength: It identified manipulation-like behavior early, allowing Sales Ops to correct region-specific forecasting issues before quarter close.

3. Asset Utilization Drift Model

Purpose: Detect subtle breakdowns in how plants reported machine uptime, downtime categories, and maintenance logs.

Why now: Plant-level definitions were aligned.

Strength: Reduced unplanned downtime by 14%, because the model caught "definition bending" before it became operational failure.

4. Backlog Integrity Model

Purpose: Flag orders whose definitions did not meet the governed "order" and "backlog" criteria.

Why now: Order-to-Cash meaning stabilized; system validations enforced prospective control.

Strength: Prevented late-quarter "pull-ins" that previously created chaos.

"These models were not powerful because they were complex.

They were powerful because the meaning behind their features was now trustworthy."

For the first time, DWL had AI that did not accelerate confusion—AI that reinforced clarity.

2. MLOps Maturity—From Manual Handling to Predictable Automation

Before The Language of Truth, Ethan and the technical teams lived in a world of stitching scripts and fixing broken pipelines manually.

Now:

2.1 Model Deployment Was Containerized

Controlled use of Azure Kubernetes Service (AKS) for scalable deployment. Mirrored for AWS using Amazon EKS for teams working in cloud-hybrid mode.

Outcome:

– Any model could scale up during peak requests and down during idle periods.

2.2 Automated CI/CD for ML—GitHub Actions + Azure ML[53] pipelines / AWS SageMaker[54] Pipelines + CodePipeline

Outcome:

– Model updates no longer required nighttime heroics.
– Re-training triggered when new governed data arrived.
– Versioning was automatic; rollbacks were instantaneous.

2.3 Feature Stores Became "Contracts"

Azure Feature Store + AWS SageMaker Feature Store

Each governed feature (e.g., "Effective Price", "Active Order", "Plant Capacity") became a reusable object.

This meant:

– "Margin" no longer had 11 definitions hiding in dashboards—it had one definition feeding every model.

2.4 Monitoring Was Continuous

Drift detection

Confidence score trends

Feature distribution shifts

Outcome:

– The system—not people—caught problems first.

3. Data Engineering Maturity—The Shift That Reduced Pipeline Burden

Prospective controls fundamentally altered the tech stack:

3.1 Transformations Moved Upstream

Before: pipelines cleaned data after it broke

After: ERP, CRM, and planning systems enforced validations before bad data existed

Example of Outcome:

---

[53] **Azure ML (Azure Machine Learning)**—A cloud service that lets you design, train, and manage machine learning models with automation and governance. Technically it is a managed Azure platform for end-to-end machine learning workflows, including data preparation, training, deployment, and monitoring. Example: Azure Machine Learning deployed a model that analyzed customer asset usage to predict renewal probability and manage backlog risk.

[54] **AWS SageMaker**—A cloud service that helps build, train, and deploy machine learning models quickly and at scale. It is fully managed AWS service that provides tools for data preparation, model training, deployment, and monitoring in production. Example: AWS SageMaker trained a model on customer order history to forecast backlog levels and optimize asset planning.

If "Order Priority" was missing, SAP forced correction before saving—
Pipeline no longer needed a "missing value fix."

3.2 Harmonization Became Native
Before: "Active Customer" and "Backlog Line" definitions lived in the
pipeline.
After: "Active Customer" and "Backlog Line" definitions now lived in the
source system.
Azure Data Factory and AWS Glue[55] now handled lighter transformations.
Outcome: Pipeline cost dropped 21%.

3.3 Memory Lake Shrunk
Once prospective logic stabilized:
fewer retrospective patches
fewer delta mismatches
fewer drift anomalies
fewer late-cycle validations

The lake became leaner, smarter, cheaper.

Stillness became strength.

4.   Governance Maturity—Machine-Aligned Meaning
Governance no longer lived in meetings or documents.
It lived in the systems and the models.

4.1 Meaning Enforcement Happened Automatically
Data Stewards approved definitions
System Rules enforced them
Models validated them
MLOps monitored them
Memory Lake traced them

This was governance without bureaucracy.

It was governance as operational physics.

4.2 Confidence Contracts Became Norm
No model could operate unless:
its features were governed
its lineage was clear
its confidence threshold was met
its decision logic was explainable

---

[55] **AWS Glue**—A serverless ETL service that extracts, transforms, and loads historical data into S3-based data lakes.
DEEPAK RANA

Executives trusted the system because the system was now accountable.

5. Cultural Maturity—Strength Without Drama
The cultural shift was striking:
No last-minute dashboards
No late-night reconciliation
No emotional escalations
No "This isn't what we expected"
No heroics required

People didn't feel less busy.
They felt less burdened.

The company learned that calm was not complacency.
Calm was competence.

And competence created predictable strength.

Reflection—When Stillness Becomes Strength
What We Observed

DWL reached its first period of organizational stillness—not by slowing down, but by eliminating rework.

Prospective controls reduced pipeline burden.
Models operated on governed data.
Systems corrected meaning before people had to.
The Memory Lake became lighter and more accurate.
Cross-functional tension eased as leaders saw consistent numbers across Sales, Finance, and Operations.

What It Means

Stillness is a competency, not an absence of motion.

When governance lives inside systems, when definitions no longer drift, and when models reinforce clarity instead of compensating for confusion, the organization becomes self-stabilizing.

The culture shifts in parallel:
less heroics, more discipline;
less urgency, more control.

Calm becomes a competitive advantage.

What will happen next

Extend stability from operations → intelligence.
Ask AI not only to predict, but to listen—to context, confidence patterns, exceptions, and organizational intent.
Develop systems that wait, observe, and request clarity before acting.
Begin the transition from purely predictive systems to contextual systems.

Aphorism

Stillness is not the end of activity—it is the beginning of precision.

Guiding Principle

Stillness is not inactivity—it is the absence of contradiction.

Law — The Stability Multiplication Law

System stability multiplies organizational speed.

Equation

$S_{speed} = k_8 \times S_{stability}$

Technical Explanation

Speed S_{speed} rises with system stability $S_{stability}$.

Fewer contradictions and fewer surprises translate directly into faster, more confident execution.

Symbols

S_{speed}: Effective decision and execution speed.

$S_{stability}$: Degree of stability (few surprises, few reversals, consistent behavior).

k_8: Positive constant linking stability to realized speed.

CHAPTER 14—The Listening Machine

When AI Learns to Wait Before Acting

By Month Nineteen, something new began to surface inside DWL—not another model, not another dashboard, but a behavior.

The machines had learned to pause.
It wasn't dramatic.
It wasn't cinematic.
But it was unmistakably disciplined.

The first example came from the Customer Churn Propensity Model.
For months, it had delivered confidence scores cleanly:
0.87—high risk
0.62—medium
0.29—low.

Simple. Predictable. Useful.

Then one morning, the model didn't return a score at all.

Instead, it flagged:
"Input feature mismatch—awaiting human validation."

Ethan stared at the log.
This had never happened before.
He traced the root cause—one of the Sales regions had begun populating "Contract Status" with a new internal shorthand that wasn't part of the governed specification. The model recognized the shift, assessed that it could not safely infer the meaning, and opted out.

Not failure.
Not hesitation.
Restraint.

And restraint was a sign of maturity.

The New Threshold of Intelligence: Context Before Action

During the next Steering Committee review, Ethan explained the behavior.
"Models can now detect when human meaning diverges from governed meaning," he said. "And when they do, they choose not to act."

Daniel leaned forward. "Why didn't this happen with earlier models?"

"Because earlier models assumed meaning," Ethan said. "Now, they check it."

Priya added, "This is the transition we've been waiting for.
Intelligence that pauses is more valuable than intelligence that predicts inaccurately."

Across the table, for the first time, the executives understood the shift:

Old AI = respond automatically.
New AI = ask before responding.

It was the digital equivalent of a leader saying:
"I need clarity before making a decision."

Wisdom wasn't arriving as a breakthrough.
It was arriving as restraint.

Why the Machines Began Waiting

It wasn't magic.
It wasn't emergent consciousness.
It was engineering.

Three shifts had occurred inside DWL's architecture:

1.   Feature contracts became hard requirements
Models no longer used default imputation when features deviated from governed values.
A "Contract Status" value outside the approved set ("Active," "Ending," "Terminated") triggered a stop.

This happened through:
–   Azure ML feature validation scripts[56]
–   AWS SageMaker Model Monitor[57]

Both enforced the contract before running inference.

---

[56] **Azure ML Feature Validation Scripts**—check that input data features meet expected rules before training or scoring models, formally defined as automated scripts in Azure Machine Learning that validate schema, ranges, and business logic to ensure reliable model performance; for example, backlog features like "Order Priority" were validated in Azure ML pipelines to block inconsistent values, while AWS SageMaker feature validation scripts performed the same checks to prevent asset forecast errors.

[57] **AWS SageMaker Model Monitor**—Tracks deployed machine learning models to detect data drift and quality issues, formally defined as a managed service in AWS SageMaker that automatically monitors input features and predictions against baseline statistics to ensure model reliability; for example, customer order predictions were monitored in SageMaker Model Monitor to flag backlog anomalies, while Azure Machine Learning's Data Drift Monitor provided a parallel check on asset usage features to maintain consistency across customer datasets.

2.   Confidence thresholds became dynamic

The Confidence Ledger[58] now tracked not just model certainty, but the integrity of the inputs.

If data quality was uncertain—even when a prediction was mathematically possible—the model refused to act.

Not because it couldn't act,
but because it shouldn't.

3.   Behavioral drift models monitored human patterns

These detectors compared:
–   timing of updates
–   field completion patterns
–   override frequency
–   the style of corrections in source systems

When human patterns changed, the system assumed intent might be shifting—and paused.

It was the first sign of a listening machine.

Across functions, the behavior spread

It wasn't just the churn model.

Within ten days:
–   The Asset Utilization Drift Model paused—maintenance logs showed a sudden increase in free-text flags in one plant. The model waited for clarification.
–   The Backlog Integrity Model paused—a new Sales Ops leader pushed for a region-specific backlog override code. The model rejected the unfamiliar value.
–   The Forecast Variability Model paused—one planner began updating weekly numbers in a pattern inconsistent with their historical behavior. The system detected "override acceleration"—a sign of possible manipulation—and halted calculation.

These were models behaving less like calculators and more like junior analysts—asking questions, requesting context, seeking alignment.

The organization wasn't just operating more precisely.

It was thinking more precisely.

---

[58] **Confidence Ledger**—is a model output log capturing decision confidence scores, defined formally as a structured record of model prediction strength; for example, the Customer Churn model logged prediction probabilities daily into Azure ML or Amazon SageMaker endpoints.

The leadership shift: delegation with boundaries

During Month Nineteen executive work session, Daniel summarized the new phase:
"We used to look for models that predicted better than people. Now we need models that know when to wait for people."

Maya added, "This is the first time AI has reduced noise instead of amplifying it."

Elena, always precise, framed it differently.
"We finally have judgment that does not panic."

Priya closed the conversation with a clarity that settled everyone.
"This is the moment AI becomes safe. Not because it's smarter, but because it's aligned."

The first Listening Contract

To formalize this maturity step, Priya and Ethan drafted DWL's first Listening Contract—a simple three-part rule:
– A model must detect when meaning shifts.
– A model must pause when meaning becomes unclear.
– A model must request human validation before proceeding.

It wasn't technology heavy.
It wasn't complex.
It was a behavioral boundary.

A boundary that aligned human and machine judgment.

And for the first time… people trusted the silence

There was no fear when a model paused.
No escalations.
No confusion.

People understood what the system was signaling:

"I heard something unusual.
Before I act, tell me what you meant."

It was the first moment DWL felt like a place where humans and machines weren't competing for authority—they were clarifying meaning together.

Stillness had become strength.
Listening had become intelligence.
Judgment had become shared.

And DWL quietly crossed into a new maturity threshold—one where AI
didn't just calculate…it listened.

The First Time the Machine Corrected the Room

This month had already shown DWL that models could pause.
But nothing prepared the organization for the moment a model would speak
first—and correct the humans.

It happened during the Monthly Commercial–Finance Alignment Review, a
meeting historically known for polite disagreement wrapped in spreadsheets. By
now, the Memory Lake fed unified inputs into the AI Forecast Variability Model,
but human expectations still anchored most decisions.

The room was full:
– Elena—running the session
– Maya—representing commercial truth
– Aaron—representing operations capacity
– Priya—governance lead
– Ethan—analytics and model oversight
– Several regional controllers and commercial directors joined virtually.

The topic: next-quarter forecast stability.

Sales had submitted their updated projection, reflecting higher renewal wins
in two regions. Finance accepted the uplift cautiously. Operations worried about
capacity in one plant.

It felt like a typical cross-functional negotiation—until the screen flickered.

A message appeared in the AI Forecast dashboard:
"Variance pattern detected—submitted forecast contradicts prior-cycle
behavior.

Awaiting clarification before acceptance."

The room froze.
It wasn't an error.
There was no red alert, no traceback, no crash.
It was a question.

It was a question.
From the model.

The Silence That Followed

"What just happened?" one regional controller asked.

Elena leaned closer. "That message didn't come from the dashboard interface. That came from the model monitor."

Ethan clicked the detail pane. A second message appeared:
"Signal divergence identified in Region 4—adjustment magnitude inconsistent with historical patterns, seasonality, and validated pipeline signals."

Maya reacted first. "That's not possible. Region 4 said they had new retention commitments."

Ethan responded carefully. "The model is saying: the magnitude of uplift submitted doesn't match the pipeline, contract stage data, or external signals we fed it."

One of the commercial directors joined the defense. "We did get verbal commitments—big accounts."

The model's message updated:
"Verbal commitments not reflected in CRM stage progression or contract records. Source misalignment suspected."

That did it.
Everyone turned to the Region 4 Sales Director on the video feed.

He unmuted slowly. "Look, we're late updating CRM because we're still negotiating. We thought we could adjust forecasts first and update the system later."

Ethan looked at him with empathy. "The model isn't saying you're wrong. It's saying the data doesn't support the uplift yet. It won't accept a forecast unless the underlying signals match."

Maya added, "The system is asking us to align our inputs before we move forward."

The room understood the meaning—the model wasn't correcting the forecast. It was correcting the behavior behind it.

The First Time AI Led Instead of Followed

This moment marked a turning point.
For the first time, the AI wasn't predicting, or flagging quality, or waiting.
It was challenging the humans.
Not aggressively.
Not authoritatively.
But logically.

Data first, narrative second.

Elena exhaled. "This is what governance as code looks like. The model isn't being difficult—it's being disciplined."

Aaron added, "For months we've been fixing upstream systems to stop automatic drift. Now the AI is doing the same for forecast logic."

Priya summarized it best:
"Alignment has shifted from being a meeting to being a behavior."

Why the Model Challenged the Room

Because of three design principles introduced in earlier:

1. Reasoning Graph Validation
The model checked not only the inputs, but whether the reasoning chain behind the inputs was coherent.

If uplift > historical pattern + pipeline strength,
the reasoning graph triggered a "context mismatch."

2. Cross-System Feature Verification
The model compared forecast changes to:
–   CRM stage progression
–   ERP contract terms
–   renewal-probability curves
–   external macro signals
If one system had not been updated, the model flagged inconsistency.

3. Human-Override Audit Path
The Confidence Ledger recorded every instance where humans overrode the model.
The model wasn't blocking the change.
It was signaling the risk of accepting it without alignment.

The Cultural Shift in Real Time

For the first time:
— Sales paused before defending a number.
— Finance paused before accepting it.
— Operations paused before planning against it.

Everyone waited—not for authority,
but for alignment.

The Region 4 Director nodded. "We'll update the CRM now. Give us two days."

Ethan refreshed the dashboard.
The model message changed:
"Awaiting updated source alignment.
Re-run forecast when CRM stages are consistent."

Not an error.
Not a rejection.
A boundary.

When Leadership Realized AI Had Become a Colleague

Later that afternoon, Daniel called Ethan and Priya into his office.
"Tell me exactly what happened," he asked.

Priya explained, "The model enforced what we agreed to:
If humans change the forecast, the system requires the source-of-truth to reflect it. No shortcuts."

Daniel nodded slowly.
"This is the first time AI prevented a misalignment before it reached the board."

Ethan added, "It's also the first time the machine protected the humans from overcommitting."

Daniel's expression softened—not because AI had become powerful,
but because DWL had become responsible enough to hear it.

He concluded:
"For the first time, we didn't negotiate truth.
We aligned it."

The Aftermath: Trust, Not Tension

The shift was immediate:
– Sales updated the CRM rigorously.
– Finance cross-verified signals before accepting uplifts.
– Operations validated capacity impacts before planning.
– The AI Forecast Model became a "neutral arbiter" everyone respected.

People stopped arguing over numbers.
They started discussing meaning.

This was the moment the Listening Machine became the Balancing Machine—discerning, steady, and aligned.
Stillness now had a voice.
And people listened.

How the Quiet Arbiter Actually Works

Although the Listening Machine felt new to DWL, every part of its behavior was built using realistic, modern enterprise capabilities:

1.   Definition Anchoring
The model checked each incoming data point against the governed definitions stored in DWL's semantic catalog (Microsoft Purview[59] for lineage + governed metadata; AWS Glue Data Catalog where applicable).
This ensured the AI used the same meaning Finance, Operations, and Sales agreed to.

2.   Cross-System Consistency Checks
The Listening Machine used lightweight validation scripts triggered in Azure Functions[60] or AWS Lambda[61] to compare updates across ERP (SAP), CRM (Salesforce), and Planning (SAP IBP[62] or Anaplan[63]) before accepting the change.

---

[59] **Microsoft Purview**—Helps organizations manage, classify, and govern their data across systems, formally defined as a unified data governance and compliance platform that catalogs, secures, and monitors information to ensure consistent policies and regulatory alignment; for example, customer order records were cataloged in Microsoft Purview to enforce compliance rules, while Azure Data Factory pipelines and AWS Glue jobs used the same governance metadata to keep backlog and asset data consistent across platforms.
[60] **Azure Functions**—Executes small pieces of code on demand without infrastructure setup, formally defined as Microsoft's serverless compute platform that responds to triggers from data, events, or APIs; for example, backlog validation was automated in Azure Functions when new customer orders arrived, while AWS Lambda provided the parallel automation for asset updates.
[61] **AWS Lambda**—Runs code automatically in response to events without managing servers, formally defined as a serverless compute service in AWS that executes functions triggered by data or system events; for example, customer order updates were processed in AWS Lambda to adjust backlog records, while Azure Functions performed the same event-driven logic for asset synchronization.
[62] **SAP Integrated Business Planning (IBP)**—Supports real-time supply chain and financial planning, formally defined as SAP's cloud solution that integrates demand, inventory, and supply planning into one framework; for example, backlog orders were balanced against asset availability in SAP IBP, with Azure Data Factory and AWS Glue pipelines ensuring consistent customer data integration.
[63] **Anaplan**—Enables connected planning across finance, supply chain, and operations, formally defined as a cloud-based planning and performance management platform that models scenarios and aligns business decisions; for example, customer demand forecasts were modeled in Anaplan to adjust backlog planning, with Azure Synapse and AWS Redshift used to store and analyze asset data feeding those forecasts.

If upstream systems didn't match, the model paused.

3. Reason Traceability

The model used SHAP (Shapley values) and Local Interpretable Model-agnostic Explanations (LIME)[64] for local interpretability, to generate human-readable reason codes.

Every forecast now came with explanation tables: which backlogs moved, which customer statuses changed, which concessions were applied.

4. Event-Driven Refresh

Whenever upstream data changed, events were published to Azure Event Grid[65] or AWS EventBridge[66].

The model retrained or revalidated only when needed.

5. Rule Layers

Simple validation rules (transactions out of range, missing contract IDs, unreasonable lead times) were implemented in Azure Data Factory or AWS Glue before the model even touched the data.

This protected the model from upstream errors automatically.

6. Listening Behavior

This wasn't sentience.

This was a combination of:

–   anomaly detection
–   lineage checks
–   policy-validated definitions
–   missing-evidence checks
–   cross-system comparison
–   SHAP-based explanation

Together, these mechanisms made the AI appear to be listening—when in reality, it was auditing alignment before generating a prediction.

In other words:

The Quiet Arbiter wasn't magic.

It was governance executed at machine speed.

---

[64] **Local Interpretable Model-agnostic Explanations (LIME)**—Explains machine learning predictions by showing which features influenced them, formally defined as an algorithm that generates local surrogate models to interpret complex ML outputs; for example, a customer churn prediction was explained with LIME in Azure ML by highlighting order cancellation features, while AWS SageMaker used LIME to clarify backlog risk classifications for asset forecasts.

[65] **Azure Event Grid**—A service that routes events between applications for real-time response, formally defined as Microsoft's fully managed event distribution platform for reliable, scalable event-driven architectures. Example, Azure Event Grid triggered a workflow when a customer order was updated, ensuring backlog records stayed consistent across asset systems.

[66] **AWS EventBridge**—A service that connects applications using events for automation and integration, formally defined as Amazon's serverless event bus for building event-driven applications across AWS and SaaS services; for example, AWS EventBridge captured customer asset changes and automatically updated backlog forecasts to maintain order accuracy.

The Day AI and Humans Disagreed

The Month Twenty began like any other in DWL's emerging stillness.
Definitions were aligned.
Upstream systems were stable.
The Memory Lake remained synchronized.
Most forecasts required only light adjustments.
The Listening Machine flagged fewer inconsistencies each week.

But stability has a paradox: it can hide the moment when a deeper test arrives.

For DWL, that moment came in early on the twentieth month.
It started with a single customer.
The issue emerged from Commercial organization—not from numbers, but from a relationship.

A major global account, Helix Industrial, had sent signals that renewal might be at risk. Helix represented nearly 6% of DWL's annual revenue, and for years, their pattern was predictable:
- long sales cycles
- large annual renewals
- stable margins
- minimal concession requests

But in late in the twentieth month, Helix made an unusual move.

They requested a significant price reduction—quietly, informally, and without the typical negotiation steps.
It hit Maya's desk before it hit any system.

Maya read the note with a frown.
"This isn't normal," Maya said.

Her commercial director answered, "They've had leadership turnover. New procurement head. They're pushing every supplier. Everyone is getting hit."

Maya didn't like this explanation, but she didn't reject it either.
"Let's at least see the numbers," she said.

The First Sign of Conflict

When Commercial submitted Helix's updated renewal probability and proposed margin reduction, the Listening Machine processed it instantly—and returned something DWL hadn't seen in months.

"Renewal-risk adjustment rejected.
Evidence inconsistent with 36-month behavioral pattern.
Margin impact exceeds allowable variance.
Reasoning trace attached."

Elena raised her eyebrows. "The model rejected the adjustment?"

Aaron leaned forward. "First time in months."

Maya stared at the message.
"We provided an explanation. Procurement change. Industry pressure."

Priya responded calmly.
"But the model doesn't use narratives as evidence."

Ethan added,
"The model learned that Helix only deviates after sequential negative indicators. This single change doesn't fit their pattern."

Maya pushed back.
"But this case isn't like its past pattern."

And there it was.

The first real tension between machine logic and human instinct.

The Governance Meeting: Humans vs. Signal

Priya called an emergency meeting with:
— Daniel
— Elena
— Maya
— Aaron
— Ethan
— and the extended governance staff

This was the first time in the entire two-year journey that the Listening Machine had rejected a strategic human judgment.

Maya opened the meeting bluntly.
"I know what the model is seeing. But this is not a standard negotiation cycle. Something has changed at the account."

Ethan replied, choosing his words carefully.
"The model isn't saying you're wrong. It's saying the evidence doesn't support a shift yet."

"Yet," Maya repeated. "And if we wait for 'yet,' we might lose the customer."

Daniel spoke next, steady and neutral.
"Let's walk through the reasoning trace."

The Model's Explanation: Cold Logic

Ethan displayed the trace on the main screen.

Reason Trace Summary:
1.   No negative usage trend (consumption stable)
2.   No support ticket surge
3.   No contract non-compliance alerts
4.   No upstream production complaints
5.   Financial stability unchanged
6.   Comparison against peer accounts: no macro trend

Then the line that created the tension:
"Current risk adjustment is inconsistent with Helix's behavioral signature."

To the model, the human input was noise.

But to Maya, it was signal.

Humans Have Context AI Doesn't

Maya leaned forward.
"The model assumes behavioral signatures don't suddenly change. Procurement leadership did change. That's exactly the kind of disruption models don't see."

Ethan nodded.
"Correct. The model uses structured evidence—system signals."

"And people use unstructured evidence," Maya said. "Tone. Silence. Pressure. Informal signals. Things you can't store in a contract file."

Priya summarized the tension perfectly.
"The model is defending definition integrity.
Maya is defending customer reality."

Daniel asked the only question that mattered.
"Is the model protecting us from overreaction…or preventing us from acting on legitimate risk?"

Everyone looked at Ethan.

He didn't hesitate.
"Both are possible."

This Was Not a Technical Problem

The disagreement wasn't about:
—   data quality
—   definitions
—   upstream systems
—   model accuracy
—   governance rules

It was about something deeper:
What happens when human intuition and machine logic disagree—and both are reasonable?

The company didn't have a protocol for this yet.

Which meant they needed to create one—carefully.

Creating the First "Human Override Protocol"

Priya led the design.

1.   Humans must provide structured evidence for override
Not rumors.
Not opinions.
Not unstructured fear.

Structured evidence meant:
—   documented stakeholder change
—   documented industry pressure
—   documented communication patterns
—   documented risk scenarios
—   documented customer history anomalies

2.   Override must be tagged as "exceptional"
Not a new rule.
Not a trend.

Just an exception.

3.   The model must incorporate override as a learning event
Not immediately.
Not blindly.

But as a data point.

4.   Override must expire after 60 days
If new signals emerged, the model could adopt them.
If not, human judgment had to revert.

Daniel approved the design.
"Machines must learn from judgment," he said.
"But judgment must also learn from machines."

Testing the Protocol

Maya provided structured evidence:
1.   Confirmation of Helix procurement leadership turnover
2.   Three competitor concessions in the same industry segment
3.   A pattern of delayed responses from the client
4.   High-value orders placed on hold
5.   Executive-level silence for 10+ days (an uncharacteristic break)

It was enough.

Priya nodded. "Override approved."

The Listening Machine updated:
"Human override accepted.
Temporary confidence adjustment applied.
Monitoring for corresponding behavioral signals."

And DWL witnessed its first moment of true human–machine collaboration.

The Aftermath: The Machine Was Right—and Wrong

Three weeks later, new evidence emerged:
—   delayed usage
—   stalled contract processing
—   a formal demand for concessions
—   competitor penetration signals
—   procurement escalation

 — abnormal account silence

In short, Helix was at risk.

The model updated itself automatically:
"Behavioral-signature shift detected.
Human override incorporated into pattern.
Risk level updated."

Maya was right.
But so was the model.

Maya brought intuition; the model brought rules.
Together, they produced the correct decision.

Why This Moment Mattered

This was not about:
— innovation
— automation
— efficiency
— forecasting
— accuracy

It was about something more fundamental:

Trust—not in AI, but in the relationship between humans and AI.

DWL had reached the threshold where:
— AI defended discipline
— Humans defended context
— Governance mediated both

And for the first time in DWL's history, decisions were stronger because disagreement was allowed.
Not suppressed.
Not overridden impulsively.
Not automated blindly.
Managed—carefully.

This was the beginning of a new maturity:
wisdom, not intelligence.

A new phase started here:

This moment—
the tension,
the override,
the correction,
the learning—became the bridge into "The Integration".
Where:
— humans and machines stop competing
— they start complementing
— context and consistency merge
— governance becomes intuition
— and DWL becomes an organization that "thinks before it predicts"

Exactly as intended from the beginning.

The Ethical Horizon.

The Month Twenty began with a clarity DWL had never experienced before.

The Helix incident had changed everything—not because it proved that machines or humans could be wrong, but because it revealed a truth the board had quietly feared:

Intelligence without boundaries is not intelligence.
It is liability.

And now, DWL had reached the line where boundaries needed to be drawn.

1. The Question No One Wanted to Ask
Daniel convened the next leadership forum—a monthly, closed-door discussion reserved for the most sensitive topics.

The attendees were familiar, steady:
— Elena—CFO
— Aaron—COO
— Maya—Head of Sales & Customer Operations
— Priya—Head of Governance & Risk
— Ethan—Strategic Analytics
— Jonas—Lead Data Engineer
— Leah—Internal Audit
— Kera—Operator / MLOps
— And, for the first time in months, Ravi—CIO (unofficially observing)

Daniel opened the session without slides, charts, or preamble.
"We've learned that AI can protect us," he said. "We've learned it can correct us. And last month, we learned it can disagree with us."

He paused.
"And now we need to decide something harder."

The room was silent.

Daniel continued.
"What happens when the system believes a decision is unethical… even when we don't?"

No one spoke.

It was the question behind the question.
The horizon they had been walking toward without naming it.

The real theme of "Earned intelligence".

Machine judgment + Human judgment = Balanced authority.

But what if the two come into conflict not just over evidence,
but over values?

2.   The First Ethical Conflict Emerges
It didn't take long for theory to become reality.

Early in Month Twenty One, the Listening Machine flagged an issue in workforce planning:

"Labor-Optimization Scenario #32 rejected.
Reason: Model predicts disproportionate impact on a vulnerable worker group.
Ethical threshold crossed.
Evidence and explanation attached."

Aaron stared at the message.
"What threshold is it using?" he asked.

Ethan replied,
"It's using the workforce-fairness rule we embedded last quarter."

Aaron frowned.
"That rule was for automated scheduling changes—not strategic workforce restructuring."

Priya intervened.
"The system doesn't know the difference unless we draw the line."

Jonas displayed the reasoning trace.

Reason Trace Summary—Ethical Flag #32
–   70% of workforce impact concentrated in one demographic
–   High correlation with seasonal shifts
–   Comparable scenarios historically mitigated through voluntary rotation
–   No documented justification for impact concentration
–   Predicted engagement-score decline > 18%
–   Predicted productivity decline > 12%
–   Alignment with DWL's ethical policy: fail

The model had done its job.
It had surfaced ethical risk.

But Aaron was right—the scenario was part of a cost-containment review, not a system-driven schedule.
"This wasn't a machine decision," Aaron said. "It was our internal review. The model is interfering."

Maya countered,
"It's flagging something we might have missed."

Elena folded her arms.
"It's raising an ethical point. But it doesn't understand business constraints."

Daniel summarized the tension.
"Ethics in a machine is easy. Context is hard."

3.   Where Does a Machine's Voice End?
The leadership spent an hour dissecting the flag.

Every angle revealed the same contradiction:
–   The model was right statistically.
–   The business context made the pattern reasonable.
–   The ethical policy didn't differentiate intent.

Priya offered the framing that broke the stalemate:
"Machines cannot understand fairness. They can only measure imbalance."

The room quieted.

She continued.
"That means this is not an ethical failure.
It's an ethical signal."

Daniel nodded.
"So, we don't silence the machine. But we don't let it veto us either."

This required something new—something they had not yet needed:

A clear boundary between AI's ethical voice
and human ethical authority.

Leah, Internal Audit, spoke for the first time.
"Then we need an Ethical Horizon."

Everything stopped.

Ethan turned to her.
"Explain."

Leah said:
"Models should detect patterns that appear unethical.
People must decide whether the pattern is unethical."

It was the simplest possible idea.
And the one DWL had been missing.

A boundary.
A protocol.
A line no machine could cross—and no human could ignore.

4.   Designing the Ethical Horizon
Priya drafted the first version of the Ethical Horizon protocol with input from Legal, Audit, HR, Strategy, and Data Science.

It included five core components:
1.   Ethical Detection (Machine layer)
AI can flag:
—   concentration of negative impact
—   bias-like pattern clusters
—   unusual risk distributions
—   cross-group imbalance
—   disproportionate outcomes

2. Ethical Interpretation (Human)

A cross-functional panel—Governance, HR, Legal, Analytics—decides:
–   whether the flagged pattern is justified
–   whether mitigation is needed
–   whether the scenario violates policy
–   whether the scenario needs redesign

3. Ethical Override Protocol

If humans choose to proceed despite a machine flag:
–   justification must be documented
–   bias mitigation plan must be added
–   impact monitoring must be scheduled

4. Ethical Post-Decision Learning

The model receives structured feedback through its governance pipeline:
–   why humans overrode
–   what was contextual
–   what was intentional
–   what was misinterpreted

This becomes training data for next cycles.

5. Ethical Sunset

All ethical overrides expire after predetermined time windows (30, 60, 90 days), forcing reevaluation.

No loopholes.

No permanent exceptions.

No "trust us."

The First Ethical Horizon Review

A week later, DWL held its first Ethical Horizon Review—the one triggered by Labor Optimization Scenario #32.

Present:
–   Daniel (CEO)
–   Priya (Governance & Risk)
–   Elena (CFO)
–   Aaron (COO)
–   HR Director
–   Legal Counsel
–   Ethan (Analytics)
–   Leah (Internal Audit)

The model replayed its explanation.

Priya summarized the case:

"The machine signals ethical risk.

Operations argues strategic necessity.
Governance sees insufficient evidence.
Audit recommends assumption validation."

Daniel asked the only question that mattered:
"Is this ethically wrong… or merely statistically imbalanced?"

HR spoke carefully.
"This scenario does not target a vulnerable group intentionally; the concentration results from seasonal staffing patterns." We can mitigate impact through rotation instead of concentrated reduction."

Legal added,
"There is no legal or compliance risk if rotation is applied."

Priya concluded,
"So the model flagged potential bias, not actual bias."

Ethan nodded.
"We should reinforce the rule to distinguish structural imbalance from temporary operational imbalance."

Daniel made the final call.
"We proceed with mitigation.
We update the model's ethical logic.
Override approved—with transparency."

The model logged:
"Human override accepted with mitigation.
Ethical Horizon Protocol engaged.
Monitoring window: 60 days."

And DWL crossed another maturity threshold.

Why This Ethical Framework Mattered in the Journey

The Listening Machine had learned to wait.
It had learned to explain.
It had learned to disagree.
And now, it had learned to defer.

But humans had learned something even more important:
AI cannot replace conscience;
it can only illuminate where conscience is required.

This was the Ethical Horizon.

Not a limit on intelligence—but the boundary that protects the organization from becoming dependent on it.

And it marked the moment DWL became something rare:

A company where, machines elevated ethics, instead of constraining humans.

Reflection—New Maturity Threshold
What We Observed
DWL reached a new threshold of maturity.
Models began to pause, question, and wait when meaning drifted.
What looked like hesitation was actually disciplined alignment.
Human–machine disagreement surfaced not as conflict, but as clarification.
And the organization learned that listening—not predicting—is the first act of intelligence.

What It Means
This was not the rise of automation.
This was the rise of accountability.
The model's restraint protected DWL from premature decisions.
The override protocol protected it from narrow interpretation.
And the Ethical Horizon defined the boundary where machine insight ends and human judgment begins.

DWL learned that intelligence is mutual:
Machines enforce consistency.
Humans enforce context.

And judgment belongs to both—but never exclusively to one.

What will happen next
DWL will extend its Listening Contracts across every model.
AI will learn to distinguish data imbalance from legitimate human context.
Ethical overrides will become structured signals for model retraining.
Governance will ensure that neither machine logic nor human intuition dominates—each will validate the other.

DWL will move toward its next maturity threshold:
systems that understand context before acting, and leaders who expect alignment before deciding.

Aphorism
Wisdom begins the moment judgment listens before it acts.

Guiding Principle
A system that knows when not to act is wiser than a system that acts blindly.

Law—The Context-Before-Action Law
Model reliability increases when the system pauses in the presence of unclear meaning.

Equation

$R_{reliability} = k_9 \times (1 - U_{uncertainty})$

Technical Explanation

Reliability $R_{reliability}$ increases as semantic uncertainty $U_{uncertainty}$ decreases.
When uncertainty rises, reliable systems pause or ask for clarification rather than acting.

Symbols

$R_{reliability}$: Reliability of model decisions in production.
$U_{uncertainty}$: Level of unresolved ambiguity in inputs or meaning (0 = clear, 1 = very unclear).
k_9: Positive constant scaling certainty into reliability.

CHAPTER 15—Truth Won

"The Architecture That Endured"

By the twenty fourth month, DWL's architecture no longer resembled the one they started with.
It wasn't bigger.
It wasn't flashier.
It wasn't a showcase of the latest cloud buzzwords.
It was simply coherent.

And that was the point.

The architecture that remained—the one that would endure beyond the transformation—had a single unifying design principle:

Every system, every dataset, every workflow must reinforce meaning, not reinterpret it.

This was the opposite of how DWL had operated for years.
Systems had been layered over systems.
Dashboards over dashboards.
Definitions drifted quietly.
Pipelines grew brittle.
ML models behaved like black boxes trying to compensate for inconsistent inputs.

What survived the two-year transformation was the architecture that supported three things:

1. A unified set of enterprise-critical data objects
Not just the original ten—but the expanded twenty-two that a real operating model depends on.

These were now classified into three layers:
    1.1. Enterprise Core Objects (the non-negotiables)
- Order
- Customer
- Backlog
- Asset
- Product / SKU / Item Master
- Vendor
- Inventory Position
- Contract / Agreement
- Work Order / Production Order

    – Invoice

1.2. Financial & Reporting Objects
- Margin
- Payment / Receipt
- Ledger Entry / Journal
- FX Rate Object
- Financial Calendar

1.3. Operational & Predictive Objects
- Forecast Unit
- Service Event
- On-Time
- Exception
- Data Quality Rule Object
- KPI Definition Object
- Risk / Control Point

This was not an academic list.
Each object was mapped systematically across:
- Order-to-Cash (O2C)
- Source-to-Pay (S2P)
- Plan-to-Build (P2B)
- Record-to-Report (RTR)
- Design-to-Insight (D2I)
- Design-to-Decision (D2D)

The architecture worked because the data objects were the architecture, not the tools.

2. A simplified, stable system landscape

Instead of optimizing every application in isolation, DWL stabilized the integrations between:
- ERP (SAP S/4HANA)
- CRM (Salesforce)
- CPQ[67] platform
- MES (Manufacturing Execution System)
- WMS (Warehouse Management System)
- Planning / APS system
- Master Data Management (MDM)

---

[67] **CPQ (Configure, Price, Quote)**—automates the process of configuring products, applying pricing rules, and generating customer quotes, formally defined as enterprise software that streamlines sales operations by embedding product logic, discount policies, and approval workflows into system processes; for example, a backlog of "Active Orders" was priced and quoted through Salesforce/SAP CPQ, with integration pipelines feeding validated quote data into Azure Data Factory and AWS Glue for downstream asset and customer reporting.

- Azure Data Lake[68] + AWS S3[69] hybrid landing zone
- Feature Store (Azure ML + AWS SageMaker parallel footprints)
- Analytics (Power BI + Tableau[70])

For the first time in years:
- The ERP did ERP work
- The CRM did CRM work
- The Data Lake did analytics work
- AI models ran on governed data objects, not stitched-together extracts

Nothing exotic.
Just clean, aligned, boringly reliable stability.

3. Prospective controls embedded directly into upstream systems
This was the breakthrough.

Instead of letting the pipeline "fix" data, DWL:
- Put validation rules in SAP for Orders, Assets, and Production Orders.
- Put customer classification rules in Salesforce.
- Put contract-obligation rules in CPQ.
- Put forecast guardrails in the APS.
- Put inventory constraints in the WMS.
- Put exception logic in MES.

The architecture became "quiet" because upstream systems were no longer producing chaos downstream.

What remained was simple:
- A lake built on meaning.
- A pipeline built on prevention.
- A model layer built on transparency.
- A governance layer embedded into the workflow.
- A cultural layer that valued clarity over speed.

It wasn't the architecture they imagined at the start.
It was better.

---

[68] **Azure Data Lake**—Stores and manages large volumes of structured and unstructured data for analytics, formally defined as Microsoft's scalable data repository designed for big data processing and integration with analytics services; for example, customer order records were ingested into Azure Data Lake for backlog analysis, while AWS S3 Data Lake served the parallel role of storing asset and customer datasets for downstream reporting.
[69] **AWS S3**—A cloud storage service that securely holds data (like revenue, orders, or customer records) and makes it accessible for analysis or applications across AWS and Azure.
[70] **Tableau**—Enables users to explore and present data through interactive charts and dashboards, formally defined as a business intelligence and analytics platform that integrates with diverse data sources to provide visual insights; for example, customer order trends were analyzed in Tableau using Azure Data Lake as input, while AWS S3 Data Lake served the parallel role for backlog and asset visualization.

The Operating Model That Stayed

End-to-end Target Operating Model (TOM)[71] structure:
Process → Data Object Ownership → People & Roles → Delivery Location
Strategy → Technology & Source Systems Mapping → Performance, OKR, and
KPI → Governance & Risk → Tax, Compliance, & Reporting

DWL's executives once thought operating models were theoretical.

By the twenty-fourth month, they understood that a modern operating model
is the only way to prevent drift from re-emerging.

The TOM they built wasn't a slide deck.

It was a living system across eight layers.

Layer 1—Mega-Process Backbone

Every activity in DWL now mapped to one of six end-to-end processes:
1. Order-to-Cash (O2C) – Sales & Distribution
2. Source-to-Pay (S2P) – Procurement
3. Plan-to-Build (P2B) – Manufacturing
4. Record-to-Report (RTR) – Financial and Management Accounting
5. Design-to-Insight (D2I) – Performance and Reporting
6. Design-to-Decision (D2D)—AI lifecycle

Each mega-process was tied to its critical data objects, creating an
unbreakable structure.

For example:

O2C:
– Order
– Customer
– Contract
– Invoice
– Payment / Receipt

P2B:
– Work Order
– SKU / Item Master
– Asset
– Inventory
– Production Event

D2D (AI):
– Forecast Unit
– Exception

---

[71] Target Operating Model (TOM)—defines how a business should operate in its future state by aligning processes, technology, and people to strategic goals, formally defined as a structured blueprint that translates business strategy into operating processes, governance, and system design; for example, a TOM ensured backlog management processes were standardized so that customer orders flowed consistently through SAP into Azure Data Factory pipelines and AWS Glue transformations, keeping asset records synchronized across platforms.

DEEPAK RANA

- Service Event
- KPI Definition
- Risk / Control Point

This was the first time DWL had true horizontal visibility.

Layer 2—Data Object Ownership

Every critical object gained:
- Primary Owner
- Secondary Owner
- Definition Steward
- Technical Steward
- Quality Owner

Before, governance was a burden;
Now, governance was a map.

Layer 3—People & Role Clarity

Roles were aligned to the new operating model:
- Global Process Owners
- Data Stewards
- Control Owners
- AI Model Stewards
- Data Engineers
- MLOps Engineers
- Domain SMEs
- FP&A Analysts
- Commercial Analysts
- Plant & Network Controllers

DWL finally answered the question every company avoids:

"Who owns the decision when systems disagree?"

Layer 4—Delivery Location Strategy

A real TOM requires clarity about where work is done:
- Complex governance → Onshore
- High-judgment modeling → Hybrid
- Standardized transformation → Nearshore
- Data pipeline operations → Offshore COE
- AI monitoring → Centralized MLOps team

This simple split improved cost, quality, and control.

Layer 5—Technology & Source Systems Mapping

"Every end-to-end process now had a clearly defined system of record:
- SAP S/4HANA—Orders, Assets, Work Orders, Invoices
- Salesforce CRM—Customers, Opportunities, Commercial Stages

- MDM Platform—Product, Customer, and Vendor Master
- APS / IBP—Forecast Units and planning signals
- MES—Production exceptions and service events
- WMS—Inventory Position and movement states
- CPQ—Contract terms and pricing logic
- Azure Data Lake + AWS S3—governed semantic layer
- Feature Store (Azure ML + SageMaker)—aligned model inputs
- Analytics Layer—Power BI and Tableau as downstream consumers"

Everything was mapped, traceable, and governed—meaning every downstream number could be explained without debate.

Layer 6—Performance, OKR, and KPI

KPI definitions became governed assets, each tied to its parent data object:
- On-Time Performance → On-Time object
- Forecast Accuracy → Forecast Unit
- Asset Reliability Index → Asset + Service Event
- Service Resolution Time → Service Event
- Margin Accuracy → Margin + Contract
- Data Quality Index → Data Quality Rule Object
- Exception Frequency → Exception
- Model Confidence Score → Risk / Control Point

Every KPI now resolved to a single, auditable meaning—no parallel versions, no localized interpretations.

Layer 7—Governance & Risk

Every mega-process operated within a governed control spine that included:
- Controls tied to specific data objects
- Standards embedded in upstream systems
- Validation rules executed before save
- Change approvals captured through workflow
- Term definitions centralized in the semantic catalog
- Drift detection across systems and models
- AI monitoring via MLOps dashboards
- Audit trails across source → pipeline → model → decision

Governance shifted from an event to an operating rhythm—the invisible scaffolding that kept behavior consistent.

Layer 8—Tax, Compliance, & Reporting

Each mega-process now carried embedded tax logic:
- Margin tied to governed cost objects and aligned with transfer-pricing policy
- Asset definitions consistent with depreciation rules across regions
- Customer definitions tied to jurisdiction, nexus, and exemption status
- Contract terms synchronized with revenue-recognition logic in CPQ

platform

Finally, the entire organization saw the same truth as Finance, Sales, Procurement, Operations, and Tax—no quiet overrides, no parallel spreadsheets, no hidden views hidden in local systems.

The Model Layer That Evolved

With the architecture stable and the operating model formalized, DWL advanced its modeling capability—slowly, deliberately, and feasibly.
This was the layer that truly transformed.

1.   The First Model Set (Aligned to the fifty Objects)
Models were no longer built from ungoverned extracts or improvised datasets. They were tied directly to governed data objects—the same fifty that now defined how DWL worked.

The core model set included:
–   Demand Forecasting Model (Inputs: Forecast Unit, Customer, SKU)
–   Service Event Prediction Model (Inputs: Asset, Service Event, Exception)
–   Exception Probability Model (Inputs: Work Order, Inventory Position, On-Time)
–   Margin Variance Model (Inputs: Contract, Invoice, Cost Object)

Each model had:
–   A clear purpose
–   Approved, governed inputs
–   Semantic alignment with the enterprise objects
–   Traceability from source to decision
–   Confidence scoring
–   Human-readable explanations

No black boxes.
No improvisation.
No "mystery features" trying to repair broken meaning.

2.   **Governing Feature Engineering**
Feature engineering stopped being a mechanical task and became a semantic one. Examples:
–   Forecast bias was derived from the governed Forecast Unit object.
–   Lead-time variance came from the governed On-Time object.
–   Service severity index was built from Service Event.
–   Asset reliability index combined Asset and Exception.

The models did not invent new meaning.

They derived features from meaning that had already been agreed, defined, and enforced.

That is why prediction error dropped from 18% to 7% in nine months.

The improvement came less from clever algorithms and more from disciplined semantics.

3.   MLOps That Stayed Simple

DWL resisted the urge to build an ornate MLOps machine.

Instead, it chose a simple, durable pattern:

– Azure ML and AWS SageMaker were used only where necessary.
– Model versioning followed a single, standard approach.
– Pipelines were intentionally minimal.
– Drift detection was straightforward and explainable.
– Retraining cycles were monthly and tied to governed data refreshes.

Maturity came from discipline, not complexity.

The system was sophisticated in logic but simple in posture.

4.   Human–Machine Integration

Every model had to behave as a colleague, not an oracle.

Models were required to surface four things:

– What they saw
– Why they saw it
– Whether humans agreed
– Whether humans should override

This was not decorative. It was the core of trust.

Executives could see the logic, challenge it, refine it, and feed that refinement back into the models.

Machine judgment and human judgment were no longer competing—they were being calibrated against the same meaning.

5.   Technical Debt Went Down, Not Up

Most AI programs quietly increase technical debt. DWL's did the opposite.

Because prospective controls and governed semantics were embedded upstream, DWL saw:

– Pipeline transformations reduced by 35–45%
– Pipeline breakage reduced by 50%
– Manual reconciliations reduced by 60%
– Model instability reduced by 30%

This is rarely discussed in AI transformations, but it was the difference between success and entropy.

DWL didn't just add intelligence.

It removed fragility.

What the Leaders Learned

Enterprise-grade, TOM-aligned, psychologically and technically integrated

By Month Twenty Fourth, DWL's leadership team finally understood something they had missed for years:

Every operational failure they had ever blamed on "systems" was really a failure of alignment.

Not alignment of personalities—alignment of meaning, ownership, and behavior.

That realization reshaped each leader in a different way.

Daniel Shaw—CEO

For most of his career,
Daniel believed strategy breaks when execution falters.
He now understood the deeper truth:
Execution breaks when meaning drifts.

His most humbling moment came when the Memory Lake produced three different "truth arcs" for a single quarter—one each for Sales, Operations, and Finance.

All valid.
All internally consistent.
All misaligned.
And all invisible to him before the transformation.

His lesson became simple:
"If leaders do not define meaning, meaning will define the business for them."

From then on, no board pack was built on assumed definitions.
Every board agenda included:
–   What changed in how we measure?
–   What changed in how we decide?
–   What changed in how the system interprets us?
He no longer feared complexity.
He feared silent drift.
That shift in fear is what turned him into a different kind of CEO.

Elena Park—CFO

Elena was transformed more than anyone.
As a finance leader, she had always treated the numbers as truth.
She now understood that numbers only reflect the quality of meaning beneath them.

She carried three lessons forward:

Lesson 1—Clarity beats precision
DWL once had extremely precise numbers that meant different things in different parts of the company. Precision without alignment is noise.

Lesson 2—Governance is a financial instrument
After prospective controls stabilized upstream systems, margin swings became smaller, predictable, and explainable.

Now she could tell the board:

"Our numbers reflect our system's behavior—not people's interpretations."

Lesson 3—Finance is a meaning function
Not just a reporting function.
Not just a policing function.
A meaning function.

Elena became a steward of semantic integrity—and came to see that stewardship as core to financial leadership.

Aaron Cole—COO
Aaron learned the most operationally painful truth:
Operations were creating 40% of their own exceptions.

Not because plants were inefficient.
Not because systems were obsolete.
But because every plant used the term "On-Time" differently.

Once prospective controls enforced a single definition:
– Exception volume dropped 28%
– Production variance dropped 14%
– Service-level instability dropped 9%

The lesson that stayed with him:
"Operational stability is not built in plants. It is built in definitions."

He also learned that upstream fixes—once seen as "technology's job"—were, in fact, his job.

Now, Operations became an active partner in data design, not a downstream victim of it.

Maya Chen—Head of Sales & Customer Operations
Maya had always believed her job was to protect customers.
She learned something more precise:

Protecting customers requires protecting meaning.

She saw how:
— Customer classification drift distorted forecasting
— Inconsistent contract terms distorted margin
— Fragmented backlog statuses distorted delivery expectations

Her core realization:
"You cannot deliver what you cannot define."

She became the strongest advocate for codifying:
— Customer segmentation
— Contract terms
— Concession rules
— Backlog stages
Now, Sales and Operations planned from the same truth.

Priya Nayar—Head of Governance & Risk
Priya learned that governance is not a function.
It is a psychological contract.
For years, she had been the fixer—chasing exceptions, cleaning errors that were not hers, repairing misalignments no one wanted to own.

After the transformation, she saw the shift:
— Meaning was owned by everyone
— Controls lived upstream
— Systems enforced alignment by default
— People trusted the outcomes because they trusted the definitions

The lesson that shaped her leadership:
"Governance succeeds when it becomes invisible."

From Month Twenty-Four onwards, she was no longer the person called to rescue the system. She was the architect of stability.

Ethan Anderson—Strategic Analytics
Ethan's lessons were quieter and more personal.
In the beginning, he saw fragments: variance here, discrepancies there, silos everywhere.
He thought the problem was analytics maturity.
He learned the deeper truth:

Analytics is a mirror, not a remedy.

He watched as:
- Meaning alignment reduced reconciliation effort by 60%
- Upstream corrections cut transformation logic by 40%
- Model's transparency, built trust across functions
- Governance codified behavior instead of patching it
- Systems begin correcting themselves instead of demanding constant rescue

His biggest realization:
"Intelligence is earned when the system behaves the way the organization believes."

That was the moment he understood the transformation wasn't about AI at all. It was about becoming a company that deserved to use AI.

His final lesson:
When meaning stabilizes, intelligence accelerates.

Leah Morales—Internal Audit

Leah learned that the strongest control in any enterprise is not a workflow or a policy. It is semantic consistency.

Every financial misstatement she had ever investigated, when traced far enough upstream, had a root cause in:
- Unclear definitions
- Inconsistent classification
- Undocumented adjustments
- Siloed interpretations

Now she had audit trails that spanned:
- Upstream validations
- Pipeline transformations
- Memory Lake lineage
- Model inputs
- Model outputs
- Decision logs

Her final insight:
"You cannot audit intelligence until you audit meaning."

The Summation of the Leadership Arc

Across Finance, Sales, Procurement, Operations, Governance, Technology, Analytics, and Audit, the collective lesson was the same:

DWL became mature not because it implemented AI, but because it became aligned enough to deserve AI.

Every leader carried forward one unifying truth:
Meaning is the first system.
Data is the second.
Intelligence is the third.
Culture is the force that connects them.

When Stillness Becomes Strength

By the end of Month Twenty-Four, DWL discovered an outcome no one had originally planned for:
The absence of noise had become a competitive advantage.
Not silence in the poetic sense—operational stillness:
The quiet that emerges when contradictions stop interrupting the workflow.

This was not calm-as-complacency.
It was calm-as-capability.

It showed up in five ways.

1.   Stability Became a Source of Speed
Once prospective controls matured and meaning stabilized upstream, the organization noticed something subtle:
  – Projects stopped pausing for data clarification.
  – Reviews stopped derailing over terminology.
  – Teams stopped spending hours reconciling numbers before they could begin real work.

A new sequence took hold:
  – Stable terms → stable data
  – Stable data → stable insights
  – Stable insights → faster decisions

Speed increased not because people moved faster,
But because the work stopped running in circles.
Less rework meant fewer restarts.
Fewer exceptions meant fewer interventions.
Fewer debates meant fewer meetings.
Executives felt it in planning cycles.
Engineers felt it in pipeline reliability.
Analysts felt it in modeling clarity.
Stillness was no longer the absence of action.
It was the removal of unnecessary action.

2. AI Models Began Learning from the Company's Stability

As upstream inconsistency declined, each new model required:
- Fewer features to compensate for broken upstream logic
- Fewer transformation layers
- Fewer "bandage columns" built to correct semantic drift
- Fewer retrains to keep up with silent definitional changes

Model training cycles shortened by nearly 40%.
Feature sets became cleaner, closer to the raw business meaning.
Most importantly, models no longer learned from people's workarounds.
They learned from the organization's intended behavior.

This changed the nature of machine intelligence at DWL:
- Models became simpler, and therefore more explainable.
- Simpler models were more stable.
- More stable models became more trusted.

Stillness became a feedback loop.
The quieter the data, the clearer the intelligence.

3. Technical Debt Declined Faster Than Anyone Predicted

When DWL reduced definition drift, standardization rippled through the stack:
- Fewer pipeline exceptions
- Fewer schema overrides
- Fewer "temporary fixes"
- Fewer unnecessary data marts
- Fewer one-off reports
- Fewer manual adjustments in Finance and Operations

The net result:
The technical debt curve inverted.
It was no longer growing faster than it could be managed.
It began shrinking organically.

The stability cascaded:
- Data engineering load reduced by ~30%
- MLOps overhead for retraining and drift detection decreased
- Cloud compute costs dropped as pipelines ran more efficiently
- Engineers spent more time improving architecture instead of repairing it

Executives noticed the budget.
Engineers noticed the burnout decline.
Everyone noticed the new clarity.

Stillness, it turned out, was cheap to maintain.
Chaos was expensive.

4. Culture Shifted from Heroism to Reliability
Before the transformation, DWL rewarded urgency:
- People who "fixed" numbers before the CFO saw them
- Analysts who reconciled in the dark before critical meetings
- Engineers who patched pipelines overnight
- Managers who saved deadlines with brute force

They weren't heroes.
They were symptoms.

By end of Month Twenty-Four, heroic work had almost disappeared—not because people stopped caring, but because the system stopped needing rescue.

The cultural center shifted:
- From fire-fighters → to system stewards
- From late-night fixes → to early alignment
- From silent adjustments → to governed meaning
- From local optimizations → to enterprise definitions

Employees no longer earned praise for rescuing broken processes.
They earned it for preventing breakage.
Stillness reshaped identity.
People felt safer, clearer, and more confident—not because the work became easy, but because the rules of the system became visible and trustworthy.

5. The Company Learned to Trust Its Own Memory
The most profound shift came when the Memory Lake—once a mirror of chaos—became a stable reflection of the business.

Executives began asking:
- "Does the lake agree with us?"
- "Does the lake show drift?"
- "What does the lake reveal about our behavior?"

The lake stopped surprising them—not because nothing went wrong, but because everything became traceable.

When something changed:
- Upstream definitions documented it
- The Reconciliation Ledger logged it
- Term owners reviewed it
- Analytics validated it

   —   AI models incorporated it

Meaning no longer drifted silently.
When meaning changed, it changed with the organization—not behind its
back.

Stillness became confidence.
Confidence became competence.
Competence became culture.

The Final Realization

By the end of the two-year journey, DWL understood a truth most
organizations learn too late:
"Stability isn't the end state.
Stability is the platform from which intelligence grows."

The company had not become perfect.
It had become predictable.

Once predictable, it became optimizable.
Once optimizable, it became intelligently adaptive.

A company that once broke under disagreement now thrived under
alignment.
A company that once chased clarity now generated it.
A company that once feared AI now partnered with it.

Stillness was not the absence of motion.
Stillness was the presence of meaning.

The Ending No One Noticed at First

There was no ribbon-cutting moment.
No all-hands celebration.
No standing ovation.

The end of DWL's two-year transformation did not arrive with ceremony.

It arrived quietly.

It arrived on a day when no one questioned the numbers in the weekly
performance deck—not because they chose not to, but because there was
nothing left to question.

That morning, the room felt different.

Ethan noticed it first.

It wasn't the charts.
It wasn't the tables.
It wasn't even the explanations.

It was the silence.

Not the uncomfortable pause before conflict.
The confident quiet that follows understanding.

For the first time in the company's history, the executive team spent more time discussing decisions than debating data.

That was the moment truth won—and almost no one realized it while it was happening.

The Two-Year Arc, Seen Clearly for the First Time

Later that afternoon, Daniel asked Ethan to summarize the full two-year journey on a single page.

Not a dashboard.
Not a strategy deck.
One page.

Ethan spent the evening reading old notes, reconciliation logs, term-governance threads, and model validation reports. He reopened the first dataset he had ever examined at DWL—from the days when "margin" had three meanings and five filters.

For an hour, he wrote nothing.

Then he began:

A company that did not trust its numbers
learned to trust its definitions,
then its data,
then its systems,
then its models,
and finally—it learned to trust itself.

He showed it to Daniel the next morning.

Daniel read it twice.

Then he said the line Ethan would remember for the rest of his life:

"Intelligence didn't transform us.
Alignment did.
Intelligence simply followed."

What the Leaders Saw in Themselves

Each member of the executive team recorded the lessons they had learned, feared, misunderstood, and eventually mastered.

Daniel Shaw—CEO

"Clarity is not a project. It is a discipline."
He realized leadership was less about fast decisions and more about shared starting points.

He saw how executives quietly reinforce misalignment when they:
— Ask for faster dashboards instead of clearer definitions
— Reward the loudest interpretation instead of the most accurate
— Treat disagreement as dissent instead of insight

He learned to ask a new question:
"Is this a data disagreement or a meaning disagreement?"

That single distinction saved hundreds of hours.

Elena Park—CFO

"Numbers have no power unless they are believed."

She had always treated accuracy as the objective.
She learned that agreement was the precondition for accuracy.
She discovered the emotional truth of governance:

People don't resist discipline.
They resist invisible discipline.

Once definitions became shared—and visible—Finance stopped being the referee and became the guide.

Aaron Cole—COO

"Operations run on rules. Good rules make the right behavior inevitable."
He learned that small inconsistencies upstream became large disruptions downstream.

Governance stopped looking like overhead and started looking like leverage.
As upstream definitions stabilized, exceptions dropped.
As exceptions dropped, throughput increased.
As throughput increased, predictability took hold.

Operations became—and remained—the most stable part of the business.

Maya Chen—Head of Sales & Customer Operations
"Customers trust consistency more than perfection."
She learned that clarity was a form of customer service.

Once "active customer" meant the same thing in CRM, support, and revenue tables, forecasts became stable enough for genuine strategic planning.

It ended a decade of recurring arguments about pipeline health.

Priya Nayar—Head of Governance & Risk
"Governance is not control; it is continuity."

She realized the greatest risk wasn't bad data—it was misaligned assumptions.

Her core insight:
"People comply with clarity.
They resist confusion."

Governance became the backbone of DWL, not because it enforced rules, but because it made rules obvious.

Kara M. Sloane—Analytics Architect
"Fixing systems doesn't scale. Fixing meaning does."

She learned that elegant pipelines crumble when definitions shift silently.

Her hardest year was the first one.
Her easiest year was the last.
Not because the workload collapsed, but because the work stopped fighting itself.

Ethan Anderson—Strategic Analytics Lead
"The only transformation that lasts is the one people understand."
He once believed his job was reconciliation.
He learned that reconciliation was the signal.
The real job was alignment.

He also admitted a truth he had never said out loud:
He didn't just want intelligent systems.
He wanted a company aligned enough to use intelligence wisely.

What the Leaders Would Avoid If They Did It Again

Each leader wrote down the mistakes they would not repeat.

Together, they formed an unintended manifesto:

1. Don't automate disagreement.
2. A model trained on misaligned definitions amplifies misalignment.
3. Don't skip definitions because "we're too busy."
4. You're never too busy to align.
5. You're always too busy to reconcile later.
6. Don't pilot AI before stabilizing meaning.
7. Pilots succeed. Scaling fails. Every time.
8. Don't assume shared language.
9. Prove it. Alignment is a measurable state, not a feeling.
10. Don't treat governance as a committee.
11. Treat it as a capability.
12. Don't build pipelines for workarounds.
13. Fix the source. Always.
14. Don't build models with features that exist only to compensate for upstream human behavior.
15. Correct the behavior.
16. Don't mistake data volume for intelligence.
17. Volume without meaning is noise.

These weren't slogans.

They were lived truths.

The Playbook DWL Will Use for Every Future Transformation

DWL captured its closing framework as:
"The Five Conditions of Earned Intelligence."

They were simple to state, hard to ignore:

1. **Meaning Before Data**

If the company cannot define a term, it cannot measure it.
If it cannot measure it, it cannot improve it.
If it cannot improve it, AI cannot learn it.

2. **Data Before Systems**

Systems reflect meaning.
Pipelines reflect systems.
Models reflect pipelines.
Order matters.

3. **Systems Before Models**

Fix upstream definitions →
then upstream systems →
then pipelines →
then models.
Not the other way around.

4. **Models Before Autonomy**

Autonomy is not granted. It is earned through explainability, self-reporting, and comparative validation against human judgment.

5. **Autonomy Before Wisdom**

Wisdom—the final condition—arises when the organization, not the machine, maintains meaning.

This framework became DWL's doctrine for future AI initiatives.

Not a checklist—a charter.

It also became the outline for DWL's next transformation.

How DWL Will Approach Every Future Initiative

The board approved a new motion:

"No AI initiative will enter design until the Five Conditions of Earned Intelligence are satisfied."

This was not a constraint.
It was a protection.
The motion established five minimum requirements:
1. Shared vocabulary
2. Stable upstream systems
3. Governed data lifecycle
4. Transparent model architecture
5. Human–machine alignment metrics

This became DWL's operating model for AI-enabled functions—a model they would continue to refine through managed optimization, not sporadic reinvention.

When Truth Won

Ethan left the office late on a Thursday.
In the elevator, he finally allowed himself to name what he had been feeling for months:
The company no longer needed him to reconcile its truth.
It was reconciling itself.
Not because it had perfect systems.
Not because it had perfect people.
Not because it had perfect models.
Because it had earned alignment.

The elevator doors opened.
The lobby was quiet.
The business was quiet.
His mind was quiet.

For the first time, he understood the transformation fully:

Truth didn't win because intelligence improved.
Intelligence improved because truth won.

He stepped outside.

DWL would change again—every company does.

But now it would change on its own terms.

The next chapter of its journey—and, perhaps, of yours—begins with a single question:

What if every transformation started with meaning?

That question closes this book.

It opens the next.

And it is the only way intelligence is ever earned.

Finally,

Guiding Principle
Alignment precedes intelligence—always.

Law — The Earned Intelligence Law
Intelligence becomes sustainable only when meaning, systems, and behavior are aligned.

Equation
$$I_{earned} = M_{meaning} \times S_{systems} \times B_{behavior}$$

Technical Explanation
Earned intelligence I_{earned} is the product of aligned meaning, stable systems, and consistent behavior.
If any of these is weak or missing, sustainable intelligence collapses.

Symbols
I_{earned}: Sustainable, "Earned" organizational intelligence.
$M_{meaning}$: Strength of shared meaning and definitions.
$S_{systems}$: Maturity and stability of supporting systems and architecture.
$B_{behavior}$: Degree of consistent human behavior aligned with the rules and meaning.

The Framework of Earned Intelligence—One-Page Summary

Across fifteen chapters, DWL's journey reveals a simple arc:

1. **Meaning First — Chapters 1–4**
   - Misaligned definitions (Chapter 1) fragment truth and make metrics untrustworthy.
   - Small definitional drift compounds into large reconciliation cost (Chapter 2).
   - Memory is only useful when governed by meaning (Chapter 3).
   - Definitions act as control systems when backed by ownership, lineage, and usage (Chapter 4).

2. **Visibility and Prevention — Chapters 5–7**
   - Reconciliation depends on seeing the full lineage, not just the outputs (Chapter 5).
   - Prospective controls reduce downstream chaos exponentially (Chapter 6).
   - Once memory is coherent, hidden patterns become visible (Chapter 7).

3. **Behavior and Integration — Chapters 8–9**
   - Behavioral drift appears before data drift (Chapter 8).
   - Integration quality is limited by the weakest assumption in the room (Chapter 9).

4. **Governance in the Fabric — Chapters 10–11**
   - Governance at the system level scales better than governance in documents (Chapter 10).
   - Stable feedback loops are the foundation of true system intelligence (Chapter 11).

5. **Trust, Stillness, and Listening — Chapters 12–14**
   - Interpretability builds trust; black boxes destroy it (Chapter 12).
   - Stability multiplies speed; stillness is the absence of contradiction, not motion (Chapter 13).
   - Models that know when to pause are more reliable than those that always act (Chapter 14).

6. **Earned Intelligence — Chapter 15**
   - Intelligence is not installed; it is earned when meaning, systems, and behavior align.
   - DWL becomes a company where truth stabilizes first—and intelligence follows.

In one line:
Meaning is the first system. Data is the second. Intelligence is the third.
Culture is the force that connects them.

Epilogue— "Beyond Athena"

DWL's buildings were quiet in the early hours, the way all well-run systems are. The peak of the transformation was behind them now—fifty definitions stabilized, pipelines corrected, models integrated, governance rewritten into muscle memory. Yet for Daniel, Priya, Ethan, and the leadership team, the stillness marked not an end, but a beginning.

Athena—the framework, not just the platform—had reached maturity. The Memory Lake was stable. Prospective controls were functioning. Feature stores were reliable. Confidence Contracts operated without escalation. The Listening Layer had matured from curiosity to judgment. And the company—after eighteen months of disciplined alignment and six months of refinement—finally held a position that many organizations never reach: a state where change did not destabilize the truth.

But the epilogue to every transformation is not about the systems that remain—it is about the questions that emerge when the work is quiet enough for people to think again.

That was the real "beyond" of Athena.

The Shift No One Predicted

Executives had expected improvement in reporting speed, variance reduction, cycle time, and audit readiness. They had expected cleaner margins, fewer reconciliations, and reduced data debt. They had expected machine learning to contribute incremental accuracy.

But the deeper shift was something else:
People began thinking together again.
Meetings no longer opened with, "Which number is right?"
Instead, they opened with, "What decision do we want to make?"
This was the real dividend of disciplined meaning:
When truth stops changing, thinking accelerates.

Senior leaders could now work on strategy instead of arguing about versions.
Regional controllers no longer maintained shadow spreadsheets.
Analysts no longer spent nights stitching data into coherence.
Engineers no longer babysat brittle pipelines.
Data Scientists and Modelers no longer fought upstream inconsistencies.
And business teams no longer feared that automation might add risk instead of removing it.

The organization, by understanding itself, had finally become scalable.

The Quiet Horizon

The team gathered in what had once been the "war room"—now converted into a "calm room." The name was informal, but the meaning was clear: this was where the company came to think, not react.

Ethan summarized the journey in a sentence that no longer surprised anyone: "Predicting is easy. Agreeing is hard."

Priya added,
"And agreeing is the foundation of earned intelligence."

Daniel closed his binder, the same way he'd done at the very beginning of the journey.
"Our next horizon is not technical," he said.
"It's structural. We must ensure that clarity scales faster than complexity."

There was no applause. No proclamation. Just aligned heads around the table—leaders who understood that the transformation was not something they had achieved, but something they needed to maintain.

This is the truth of all transformations:
Maturity is not a destination.
It is the discipline of guarding what was earned.

The Real Legacy of Athena

DWL did not become a company defined by AI.
It became a company defined by responsibility.
Systems corrected themselves, but only because people had corrected their behaviors.
Pipelines remained stable, but only because meaning no longer drifted.
Models improved, but only because the data afforded them that right.
Governance scaled, but only because leaders treated it as their job, not someone else's.

The real legacy of Athena was simple:
A company learned to think before it calculated.
And to understand before it automated.

What Comes After Stillness

Every transformation ends with a question:
"What do we do now that we are no longer fixing the past?"

For DWL, the answer was clear.

They would turn their maturity outward—toward suppliers, customers, partners, and eventually, toward entirely new business models. They would expand the Listening Layer beyond internal signals into ecosystem signals. They would refine Confidence Contracts into industry-level trust standards. They would evolve from a company managing its data to a company orchestrating its intelligence.

The next evolution was not about Athena.

It was about everything Athena had made possible.

This is where the story ends for now—with a company no longer defined by fragmentation,
  but by the discipline of coherence;
  no longer driven by reconciliation,
  but by informed choice;
  no longer pulled by the past,
  but ready for whatever future intelligence demands.

And somewhere in the quiet of that still organization,
  a new question waited—the kind that begins a second chapter of evolution.

Because once intelligence is earned, the real work begins:
  deciding how wisely to use it.

APPENDIX

Executive Cheat Sheet

The 10 Signals of a Healthy Data & AI Organization

1. Shared definitions across systems
2. Stable upstream data
3. Minimal pipeline transformation
4. Governance embedded as code
5. Clear decision rights with machines
6. Explanation layers on all models
7. Drift detection on meaning + data
8. Real-time reconciliation visibility
9. Trusted semantic alignment score
10. AI used only where meaning is stable

The 10 Signals of Decline (The EPAC Warning System)

1. Silent definition drift
2. Increasing pipeline patches
3. Multiple truths in the same meeting
4. Growing reconciliation debt
5. Human overrides with no trace
6. ML models losing confidence
7. New data categories appearing at source
8. Manual batching in operational steps
9. Conflicting KPIs across functions
10. "Quick wins" that expose deeper issues

Laws, principles, and formulas of Earned Intelligence

Laws of Earned Intelligence

(The 12 foundational laws that emerged from DWL's journey)

Law 1 — The Law of First Meaning
If people do not agree on meaning, systems will not agree on output.
Every transformation must begin with shared definitions; otherwise, every model amplifies inconsistency.

Law 2 — The Law of Retrospective Truth
You cannot fix tomorrow's intelligence if yesterday's data is still broken.
Retrospective reconciliation is a prerequisite for trustworthy AI.

Law 3 — The Law of Prospective Integrity
The only sustainable transformation is one that prevents new errors at the source.
Controls embedded in systems and processes remove the need for endless data cleansing.

Law 4 — The Law of Semantic Drift
All definitions degrade unless actively governed.
Language moves faster than policy; governance prevents meaning from diverging.

Law 5 — The Law of Signal Stability
Models cannot outperform the stability of the signals feeding them.
Stable pipelines → stable models; unstable signals → expensive failures.

Law 6 — The Law of Alignment Ratio
An aligned organization outperforms a technically superior one.
Cognitive alignment multiplies the impact of technical investment.

Law 7 — The Law of Human Primacy
AI can assist decisions, but only humans can contextualize them.
Every system must retain mechanisms for interpretation, correction, and ethical override.

Law 8 — The Law of Controlled Autonomy
Autonomy must be earned, not granted.
Models receive decision rights only when trust thresholds and confidence contracts are satisfied.

Law 9 — The Law of Drift Velocity
The faster the business changes, the faster models must adapt—or fail.
Monitoring drift is not optional; it is a survival mechanism.

Law 10 — The Law of Governance Saturation
Too little governance creates chaos; too much governance destroys velocity.
Balance determines maturity.

Law 11 — The Law of Institutional Calm
When systems behave predictably, teams regain strategic attention.
Operational calm is a leading indicator of organizational health.

Law 12 — The Law of Earned Intelligence
Intelligence is not built; it is earned—through meaning, consistency, oversight, and humility.
This is the transformation destination.

Principles of Semantic & Operational Alignment

(Practical guidelines used by DWL to operationalize the laws)

Principle 1 — Define Once, Use Everywhere
Definitions must be universal across all dashboards, models, and workflows.

Principle 2 — Meaning Precedes Measurement
A metric without a shared definition becomes an opinion with formatting.

Principle 3 — Correction Before Automation
Never automate a broken process; automate only after retrospective alignment.

Principle 4 — Contextual Supervision
Models must present reasoning in human language, enabling contextual validation.

Principle 5 — Global Term Ownership
Every term requires a steward, an approver, and a consumer group.

Principle 6 — Drift is a Symptom, Not the Problem
Drift signals underlying changes in behavior, process, or environment.

Principle 7 — Trust Requires Traceability
If you cannot trace how a value or prediction was produced, you cannot trust it.

Principle 8 — Prevention Outperforms Detection
Prospective system rules are exponentially cheaper than retrospective fixes.

Principle 9 — Calm is a Competency
Teams that operate without noise outperform those reacting to constant exceptions.

Principle 10 — Alignment is Everyone's Job
Data quality is not a function; it is a discipline across the organization.

Transformation Formulas (Earned Intelligence Mathematics)

Each formula is intentionally simple, allowing readers to communicate maturity in measurable terms.

1. **Formula 1 — Semantic Alignment Score (SAS)**
   SAS = (Aligned Definitions ÷ Total Definitions) × 100
   A measure of how consistently systems and teams interpret core business concepts.

2. **Formula 2 — Drift Index (DI)**
   DI = (Observed Change – Expected Change) ÷ Expected Change
   Identifies how far data or model behavior deviates from expectations.

3. **Formula 3 — Noise Ratio (NR)**
   NR = Noisy Signals ÷ Total Signals
   Quantifies how much of the environment produces confusion or inconsistency.

4. **Formula 4 — Stability Index (SI)**
   SI = Stable Outputs ÷ Total Outputs
   Higher values indicate predictable operations and reliable pipelines.

5. **Formula 5 — Trust Threshold (TT)**
   TT = Minimum Confidence Score Required for Autonomy
   The boundary a model must cross before decisions can be delegated.

6. **Formula 6 — Confidence Contract Boundary (CCB)**
   CCB = Model Confidence – Required Threshold
   Used to determine human intervention needs.

7. **Formula 7 — Governance Saturation Score (GSS)**
   GSS = Controls Implemented ÷ Controls Required
   Measures under- or over-governance.

8. **Formula 8 — Reconciliation Debt (RD)**
   RD = Outstanding Data Issues × Days Outstanding
   Indicates how much unresolved data error accumulates over time.

9. **Formula 9 — Stability Ratio (SR)**
   SR = (Calm Days ÷ Total Days)
   A macro indicator of institutional maturity.

10. **Formula 10 — Wisdom Metric (WM)**
    WM = (Explainability Quality + Context Match + Ethical Alignment) ÷ 3
    Evaluates whether a model's reasoning—not just its prediction—is fit for organizational use.

Executive Summary Sheets (Quick Reference format samples)
Sheet A — The Alignment Sheet
- Define meaning before measurement
- Reduce reconciliation load
- Ensure coherence across regions
- Monitor semantic drift weekly
- Maintain the Alignment Ratio above 90%

Sheet B — The Control Sheet
- Implement prospective controls
- Validate retrospective corrections
- Use governance-as-code for scale
- Target Control Saturation between 70–85%

Sheet C — The Drift & Stability Sheet
- Monitor Drift Index
- Track Noise Ratio
- Ensure Stability Index remains above 85%
- Investigate any sudden volatility signals

Sheet D — The Wisdom Sheet
- Require explainability for all automated outputs
- Maintain Confidence Contracts
- Review ethical overrides monthly
- Establish a Wisdom Threshold for high-impact decisions

Sheet E — The Transformation Rhythm Sheet
- Operate in 30-day governance cycles
- Recalibrate definitions quarterly
- Audit autonomy assignments annually
- Maintain Institutional Calm as a key performance indicator

Sample Implementation Blueprint (Condensed)

(A one-page transformation roadmap)

Phase 1 — Meaning (Months 1–3)
– Define core terms
– Build alignment council
– Introduce Alignment Ratio

Phase 2 — Retrospective Truth (Months 3–6)
– Clean historical data
– Deploy reconciliation ledger
– Measure Reconciliation Debt

Phase 3 — Prospective Control (Months 6–12)
– Implement system guardrails
– Introduce governance-as-code
– Track Stability Index

Phase 4 — Operational Rhythm (Months 12–18)
– Establish Athena Cycles
– Monitor drift, noise, stability
– Build Memory Lake dependencies

Phase 5 — Earned Intelligence (Months 18–24)
– Deploy explainability and wisdom metrics
– Formalize autonomy boundaries
– Evaluate Wisdom Threshold
– Sustain Institutional Calm

About the Author

Journey over time

Deepak Rana is a transformation leader with more than twenty-five years of experience guiding global enterprises through their most critical inflection points. Over his career, he has supported more than fifteen major organizations across eight industries, often stepping into programs stalled by politics, misalignment, or competing versions of truth. His hallmark is turning complexity into clarity.

He works at the intersection of business strategy, transformation, governance, operating model design, operational excellence, and executive decision enablement. In environments defined by ambiguity, Deepak is recognized for restoring alignment. He helps leadership teams rebuild trust, redefine foundational terms, and make decisions that endure.

What sets his perspective apart is lived experience. Time and again, he has been asked to resolve challenges others avoid—initiatives caught between business and technology, teams divided by incentives, and organizations overwhelmed by complexity. Through these moments, he discovered that transformation is psychological before it is technical.

A Wharton alumnus and graduate of Harvard's Master of Management program, Deepak writes to make transformation readable, relatable, and replicable for leaders at every level.

Earned Intelligence is his first narrative work, combining the precision of an engineer with the perspective of a storyteller. He lives by the rhythm he teaches—observe, interpret, act—and continues to mentor the next generation of data-literate, ethically grounded professionals who will shape the future of intelligent enterprises.

This book distills lessons from his journeys into a narrative that leaders and technical teams can use to build organizations where truth is agreed, intelligence is earned, and trust becomes a capability rather than a wish. He writes as someone who has seen the consequences of misalignment—and the power of clarity when it finally emerges.

Authorial Synthesis

How these ideas shape this book

This book is built on a simple observation: organizations rarely fail because they lack intelligence. They fail because they lack shared meaning. The disciplines that shaped this book—data governance, organizational psychology, machine learning, decision hygiene, systems architecture, operating model design, and narrative clarity—offered a unified lens for understanding that the hardest problems in enterprises are semantic before they are technical.

From leadership science came the insight that incentives and culture often determine the fate of technology long before the first model is deployed. From data governance came the understanding that definitions are decisions, and decisions become behavior. From ML and AI came the recognition that systems amplify whatever they inherit—clarity or chaos. And from narrative craft came the ability to convey these complex ideas in a way leaders and practitioners can absorb and act on.

The Matrix-archetype influence is never thematic or explicit; it is structural. The book uses the psychological rhythm of awakening: seeing a system as it is, learning its contradictions, aligning meaning, building capability, and ultimately operating with wisdom rather than reaction.

These ideas shape Earned Intelligence into a story about becoming deliberate—individually, organizationally, and technologically. It is not fiction, nor pure nonfiction, but a form designed to make the deepest truths easy to see.

Acknowledgement

This book was shaped by many people whose influence is woven through its pages.

To the executives, operators, analysts, data stewards, architects, controllers, commercial leaders, and governance teams I have worked with over the years— thank you. Every transformation, every late-night working session, every difficult reconciliation, every turning point in alignment contributed to the insights in this book. While no company or individual is represented directly, the lessons come from real struggles and real breakthroughs.

To the colleagues who challenged my thinking, opened new doors of perspective, and trusted me with complex programs—your partnership made the work meaningful.

To the mentors who believed in the power of clarity before technology— your guidance shaped the foundation of this book.

To my family— Jai Singh Rana (Father), Gurbachan Rana (Mother), Smita Rana (Wife), and Ishaan Rana (Child)—thank you for your patience, encouragement, and unwavering support through years of demanding work. Every page exists because of the space you created.

Finally, thank you to the leaders and teams who continue striving to build organizations anchored in truth rather than noise. This book is dedicated to your courage.

Reference Material

Further Reading & Influences

This book was written at the intersection of enterprise transformation, AI maturity, organizational psychology, and decision governance. The following works—academic, technical, managerial, and literary—shaped the thinking behind Earned Intelligence. They are not sources of quoted material, but foundational influences that help readers explore the space more deeply.

I.   *Enterprise Transformation, Data & Governance*

1. *John Ladley—Data Governance: How to Design, Deploy, and Sustain an Effective Data Governance Program—A foundational text on enterprise governance design and operating models.*

2. *DAMA International—The DAMA Guide to the Data Management Body of Knowledge (DAMA-DMBOK)—Canonical reference for data architecture, quality, stewardship, and lifecycle practices.*

3. *APQC (American Productivity & Quality Center)—Global standards for process classification, benchmarking, and business capability models.*

4. *Thomas Davenport—Competing on Analytics & Analytics at Work—Seminal work linking analytics capability with business performance.*

5. *Gene Kim, Jez Humble, Patrick Debois, John Willis—The DevOps Handbook—Relevant for the governance-as-code and pipeline stability themes introduced in this book.*

II.   *Cloud, Engineering, and AI/ML Platforms*

6. *AWS Well-Architected Framework (Amazon Web Services)—Industry-standard principles for secure, reliable, efficient cloud system design.*

7. *Microsoft Azure Architecture Center—Best practices for data pipelines, orchestration, and ML deployment on Azure.*

8. *Google Cloud Architecture Framework—Useful for readers wanting to compare multi-cloud AI approaches.*

9. *Martin Kleppmann—Designing Data-Intensive Applications—The definitive work on data pipelines, streaming, storage, and system reliability.*

III.   *AI, Machine Learning, Forecasting & Causality*

(Academic References All Serious AI Books Should Acknowledge)

10. *Christopher M. Bishop—Pattern Recognition and Machine Learning—The graduate-level foundation of ML theory.*

11. *Ian Goodfellow, Yoshua Bengio, Aaron Courville—Deep Learning—Seminal text for modern neural network architectures.*

12. *Trevor Hastie, Robert Tibshirani, Jerome Friedman—The Elements of Statistical Learning—Core reference for supervised learning, regularization, and feature modeling.*

13. *Andrew Ng—Machine Learning Yearning—Practical strategic guidance on ML project design and error analysis.*

14. *Judea Pearl—The Book of Why (Causality)—Defines causal reasoning and why explainability matters as much as prediction.*

15. *Rob J. Hyndman & George Athanasopoulos—Forecasting: Principles and Practice—Essential for time-series, trend analysis, seasonality, and temporal validation.*

16. *Tom Mitchell—Machine Learning—Classic introduction to ML concepts, generalization, and model behavior.*

IV. *Organizational Leadership, Psychology, and Decision-Making*

17. *Daniel Kahneman—Thinking, Fast and Slow—Central influence behind cognitive bias discussions and decision hygiene.*

18. *Chip Heath & Dan Heath—Switch (Behavior Change)—Relevance to cultural transformation themes.*

19. *Amy Edmondson—The Fearless Organization—Psychological safety framing for governance maturity.*

20. *Edgar Schein—Organizational Culture and Leadership—Mapping culture shifts to enterprise transformation.*

21. *Yuval Noah Harari—Sapiens and Homo Deus—Macro-pattern thinking underlying enterprise evolution.*

22. *Robert Cialdini—Influence and Pre-Suasion—Decision psychology principles reflected in stakeholder alignment.*

V. *Writing Craft, Narrative Flow & Authorial Technique*

(These books influenced tone, clarity, and narrative discipline—NOT content.)
23. *Stephen King—On Writing—Sentential rhythm, pacing, and micro-clarity.*

24. *Jonah Berger—Magic Words—Precision in verbal framing and communication.*

25. *James W. Pennebaker—The Secret Life of Pronouns—Insights on authentic voice and narrative realism.*

26. *Malcolm Gladwell—The Tipping Point & Outliers—Structuring inflection points and narrative transitions.*

27. *Clayton Christensen—The Innovator's Dilemma & Competing Against Luck—Jobs-to-be-done thinking embedded in organizational analysis.*

28. *Adam Grant—Think Again—Cognitive flexibility themes reflected in the maturity*

arc.

29. *Jordan Peterson—12 Rules for Life—Order–chaos framing relevant to governance stabilization.*

30. *Jeffrey Pfeffer—Leadership BS—Realpolitik framing for enterprise incentives and power structures.*

VI. ERP, Operating Model & Transformation (Non-Vendor-Specific)

(No brand-specific ERP materials are cited to avoid legal entanglement, but these categories influence the thinking.)

31. *Enterprise Operating Model frameworks—Principles used across Fortune 500 transformations: capability mapping, process architecture, governance.*

32. *Finance Transformation / FP&A Improvement Literature—Used for structuring cross-functional alignment and metric governance.*

33. *Supply Chain Operations Reference Model (SCOR)—Helpful conceptual influence for flow-based architecture (order → fulfillment → financial realization).*

34. *Management Accounting & Costing Frameworks—For structuring reconciliations, drivers, and value attribution across functions.*

Glossary

A

Accountability Layer
The combination of governance, controls, and validation mechanisms that ensure decisions reflect agreed meaning rather than local interpretation.

Active Customer (Governed Definition)
A customer classified using a unified, cross-system rule set that determines eligibility for forecasting, margin analysis, and renewal status.

AI Confidence Ledger
A structured audit trail that logs: model output, input features, confidence scores, overrides, and human reasoning. Used to track alignment between machine and human judgment.

AI Drift Monitor (Model Monitor)
A system that detects deviations in model inputs, behavior, or performance over time. At DWL, implemented with AWS SageMaker Model Monitor and Azure ML drift detectors.

AI Explainability Panel
A human-readable interface showing each feature's contribution to a model decision (SHAP, LIME), enabling leaders to understand why predictions were made.

AI Listening Contract
The rule set stating that a model must pause, request clarification, or defer when meaning is unclear or when upstream systems send inconsistent signals.

AI Pause Condition
A governed rule specifying the conditions under which a model must stop, signal uncertainty, and ask for clarification before issuing a prediction.

AI Reasoning Graph
A structured representation of how the model evaluates inputs, relationships, and decision paths—used to detect inconsistencies in human-supplied narratives.

Alignment Council
A cross-functional group responsible for maintaining shared definitions, reconciling conflicts, and preventing semantic drift across regions and systems.

Anaplan
Tool that enables connected planning across finance, supply chain, and

operations, formally defined as a cloud-based planning and performance management platform that models scenarios and aligns business decisions; for example, customer demand forecasts were modeled in Anaplan to adjust backlog planning, with Azure Synapse and AWS Redshift used to store and analyze asset data feeding those forecasts

Anomaly Detection Model
A model that identifies unusual patterns indicating data errors, system failures, fraud, or process deviations.

Athena Framework
DWL's governance architecture combining definitions, controls, model oversight, and organizational learning into a unified operational system.

AWS Glue
A managed service for ETL operations that prepares data for analytics and ML workloads.

AWS Glue Data Catalog
A central index that stores and organizes metadata about data assets so they can be easily discovered and used. Technically it is a fully managed metadata repository in AWS that enables consistent data definitions, schema management, and lineage tracking across services. Example: The AWS Glue Data Catalog organized customer order tables, allowing backlog forecasts to be run consistently across multiple analytics tools.

AWS Lambda
Runs code automatically in response to events without managing servers, formally defined as a serverless compute service in AWS that executes functions triggered by data or system events; for example, customer order updates were processed in AWS Lambda to adjust backlog records, while Azure Functions performed the same event-driven logic for asset synchronization.

AWS S3
A cloud storage service that securely holds data (like revenue, orders, or customer records) and makes it accessible for analysis or applications across AWS and Azure.

AWS SageMaker
A cloud service that helps build, train, and deploy machine learning models quickly and at scale. It is fully managed AWS service that provides tools for data preparation, model training, deployment, and monitoring in production. Example: AWS SageMaker trained a model on customer order history to forecast backlog levels and optimize asset planning.

AWS SageMaker Model Monitor

Tracks deployed machine learning models to detect data drift and quality issues, formally defined as a managed service in AWS SageMaker that automatically monitors input features and predictions against baseline statistics to ensure model reliability; for example, customer order predictions were monitored in SageMaker Model Monitor to flag backlog anomalies, while Azure Machine Learning's Data Drift Monitor provided a parallel check on asset usage features to maintain consistency across customer datasets.

Azure Data Catalog

A service that registers, classifies, and governs metadata for assets across Azure and other sources. Technically it is a metadata catalog in Azure that provides data discovery, schema definitions, and lineage tracking for compliance and analytics. Example: Azure Purview cataloged asset records and customer backlog data, ensuring consistent schema definitions for reporting and forecasting.

Azure Data Factory (ADF)

A pipeline orchestration service that moves, transforms, and validates data across systems.

Azure Data Lake

Stores and manages large volumes of structured and unstructured data for analytics, formally defined as Microsoft's scalable data repository designed for big data processing and integration with analytics services; for example, customer order records were ingested into Azure Data Lake for backlog analysis, while AWS S3 Data Lake served the parallel role of storing asset and customer datasets for downstream reporting.

Azure Event Grid

A service that routes events between applications for real-time response, formally defined as Microsoft's fully managed event distribution platform for reliable, scalable event-driven architectures. Example, Azure Event Grid triggered a workflow when a customer order was updated, ensuring backlog records stayed consistent across asset systems.

Azure Functions

Executes small pieces of code on demand without infrastructure setup, formally defined as Microsoft's serverless compute platform that responds to triggers from data, events, or APIs; for example, backlog validation was automated in Azure Functions when new customer orders arrived, while AWS Lambda provided the parallel automation for asset updates.

Azure Machine Learning (Azure ML)

A cloud service that lets you design, train, and manage machine learning models with automation and governance. Technically it is a managed Azure platform for end-to-end machine learning workflows, including data preparation, training, deployment, and monitoring. Example: Azure Machine Learning deployed a model that analyzed customer asset usage to predict renewal probability and manage backlog risk.

B

Backlog Integrity

A governed audit of backlog status consistency across systems (ERP, CRM, APS), ensuring delivery expectations match reality.

Behavioral Drift

A change in human patterns (timing, overrides, shortcuts) that precedes and predicts data drift.

Breakage Rate (Pipeline)

The frequency of pipeline failures due to schema mismatch, definitional drift, or inconsistent upstream logic.

Business Steward

The business-side owner of a term, process, or dataset, accountable for meaning, quality, and usage across the enterprise.

Bronze-Silver-Gold Tiers

Structure raw, cleaned, and governed data layers; formally defined as a three-layer data architecture (raw , cleansed, semantic) used to ensure progressive reliability. Example: In AWS S3/Athena and Azure Data Lake/Fabric, Backlog Quantity, entered Bronze as-is reached Silver after deduplication and Gold after definition alignment

C

Central Feature Store

A single repository for aligned, governed, versioned features used across all models. DWL used both Azure ML and AWS SageMaker Feature Store.

Change Saturation

The threshold at which teams experience diminishing returns from new processes or controls, used to time the introduction of governance changes.

Churn Propensity Model

A predictive model estimating the likelihood of losing a customer. At DWL, it became the first model to pause because of semantic mismatch.

CloudWatch / Azure Monitor
Systems used to monitor drift, performance, anomalies, and operational events across model environments.

Cognitive Load (Organizational)
The mental burden caused by inconsistent definitions, scattered truth, or contradictory data — reduced through alignment and governance.

Commercial Truth Arc
The historical narrative constructed by Sales data (e.g., concessions, renewals, pipeline movements) as stored in the Memory Lake.

Confidence Contract
A formal agreement between humans and AI requiring: review of explanations, documented overrides, structured disagreement, and continuous learning.

Containers (Docker / Kubernetes)
Unitized compute environments used to run models, pipelines, and microservices reliably. DWL used:
– AKS (Azure Kubernetes Service)
– EKS (AWS Elastic Kubernetes Service)

Contract Object (Governed)
A unified representation of contract terms, pricing, obligations, which removes inconsistencies between CPQ, CRM, and ERP.

Control Mesh
The collection of embedded rules, validations, and guardrails designed to prevent errors at the source before data enters analytical or AI pipelines.

Cross-System Signal Verification
A process checking whether updates in one system (CRM) match supporting changes in another (ERP, CPQ, APS) before updating forecasts or models.

D

Data Contract (Upstream)
A governed agreement specifying what fields mean, who owns them, their allowed values, and what breaks if the contract is violated.

Data Quality Rule Object

A governed rule controlling validation logic for upstream systems (SAP, Salesforce, APS). Stored in DWL's semantic catalog.

Data-to-Decision (D2D)
The end-to-end mega-process for AI lifecycle: meaning → data → system → model → decision → feedback.

Decision Drift
When different functions interpret the same scenario differently because of semantic inconsistency.

Decision Rights
The agreed ownership structure defining who can approve definitions, adjust models, or override automated recommendations.

Definition Steward
The owner responsible for maintaining the meaning, lineage, and integrity of a governed term.

Design-to-Insight (D2I)
The enterprise process that defines how insights (dashboards, reports, analytics) are produced from governed data objects.

Drift Event
Any unexpected change in behavior, meaning, or input pattern that causes a model or pipeline to pause.

Drift Indicator
A numeric or binary signal used to show whether a feature or definition is deviating from its governed value.

E

Earned Intelligence
The condition in which AI becomes sustainable because meaning, systems, and behaviors are aligned. Intelligence is earned, not assumed.

Ethical Horizon
The boundary where AI's ability to signal potential ethical issues ends and human authority begins. Machines detect patterns; humans decide intent.

Ethical Override Protocol
The governed process required when human judgment contradicts machine-recommended actions in ethically sensitive decisions.

Exception Probability Model
A predictive model estimating the likelihood of operational exceptions (late orders, bottlenecks, delays).

F

Feature Contract
A governed specification for each model feature: owner, definition, allowed values, transformations, drift rules, and version history.

Feature Engineering by Governance
The principle that features must reflect aligned meaning — not expedient computation. The foundation of DWL's stable models.

Feature Passport
A complete documentation artifact for each feature: semantics, lineage, transformations, constraints, owners.

Feature Store
Central repository for consistent model inputs; formally defined as a governed store of features used for training and inference. Example: "Customer Recency" and "Asset Downtime Ratio" were published into AWS SageMaker Feature Store and Azure ML Feature Store for standardized reuse.

Financial Consistency Index
Measures how reliably financial data aligns across systems and time, formally defined as a governance metric that validates the accuracy, completeness, and synchronization of financial transactions across distributed platforms; for example, backlog valuation was checked using a Financial Consistency Index in Azure Synapse Analytics to ensure customer order totals matched asset records, while AWS Redshift applied the same index to confirm revenue consistency across backlog and customer datasets.

Forecast Unit (Governed Object)
The core predictive object representing demand signals, tied to SKU, contract, and customer dimensions.

Full Lineage Trace
A complete record of how data moves from source to report, used for reconciliation and audit.

G

Governance as Code
Embedding governance rules directly into systems, pipelines, and models so alignment happens automatically, not through manual review.

Governance Ledger
A centralized record of every definition change, data correction, model override, and control update—ensuring traceability across time.

Governed Memory Object
Any data object that is semantically defined, lineage-validated, and behavior-aligned in the Memory Lake.

Guided Override
A controlled override where humans must supply structured evidence before contradicting a model recommendation.

H

Human–Machine Alignment Score
A metric comparing human decisions with model reasoning to detect gaps in judgment consistency.

Human Override Path
A governed decision trail showing when, why, and how humans overrode model output.

I

Inference Guardrail
A control preventing models from producing predictions with insufficient meaning clarity.

Ingest Pipeline (ADF / Glue)
The orchestrated data movement mechanism from upstream systems into the Memory Lake.

Integration Workshop
A structured cross-functional session exposing assumption drift, definition conflicts, and system inconsistencies.

K

KPI Definition Object
A governed, semantically clear metric object linking KPIs to underlying data

structures and definitions.

Kubernetes (AKS / EKS)
The container orchestration platform used to run scalable, resilient AI and data workloads.

L

Language of Truth
The initiative to align definitions, term usage, and decision logic across the enterprise.

LIME (Local Interpretable Model Explanations)
A technique used to generate human-readable explanations for individual predictions.

Listening Machine
The AI capability that pauses, defers, or escalates when meaning becomes unclear — signaling semantic uncertainty.

Lineage Object
A governed representation of data source → transformation → destination.

Logistic Regression
Estimates the probability of an event happening, formally defined as a statistical model that applies a logistic function to classify outcomes into binary categories; for example, a Logistic Regression model in Azure ML calculated the likelihood of a customer cancelling an order, while AWS SageMaker used parallel logistic regression to flag backlog items most likely to miss shipment deadlines.

M

Margin Integrity
A unified, cross-functional definition of margin enforced across Finance, Operations, and Commercial.

Meaning Drift
The silent shift in how a term is interpreted between teams, systems, or processes.

Memory Lake
The governed, lineage-tracked, semantically unified data platform replacing fragmented legacy stores.

MLOps (Machine Learning Operations)
The combined process of model training, deployment, monitoring, drift detection, and lifecycle management.

Model Workspace
A controlled environment for training, validating, and testing models (SageMaker, Azure ML).

N

Noise Reduction Index
A measure of how much downstream noise was prevented by strengthening upstream controls.

O

Object Owner
The designated steward responsible for definition, quality, lineage, and governance of a core data object (Order, Customer, Forecast Unit, etc.).

Operational Drift
Unexpected changes in plant, inventory, network, or production behavior that reveal systemic inconsistency.

Override Acceleration
A governance signal triggered when humans begin overriding models at unusual frequency or velocity.

P

Pattern Emergence Score
A metric indicating how much new insight becomes visible when data coherence increases.

Preventive Control
A rule preventing errors early (SAP validation, CRM restrictions, contract guardrails) rather than correcting them later.

Prospective Control Layer
The embedded set of rules inside upstream systems designed to prevent drift before it occurs.

Q

Quiet Arbiter
The AI capability that evaluates signals, requests clarity, enforces governance, and refuses inconsistencies — without asserting authority.

R

Random Forest
Predicts outcomes by combining many decision trees, formally defined as an ensemble machine learning algorithm that aggregates multiple tree models to improve accuracy and reduce overfitting; for example, customer churn was predicted by a Random Forest model in Azure ML using order history features, while AWS SageMaker applied the same technique to backlog and asset usage data for risk classification.

Reconciliation Ledger
A complete audit log of semantic inconsistencies, mapping mismatches, and corrections across systems.

Reasoning Trace
A step-by-step explanation of how the model reached its decision — used in interpretability.

S

Semantic Alignment
The condition in which terms have consistent meaning across systems, teams, and processes.

Semantic Catalog (Purview / Atlas)
The enterprise metadata system governing definitions, lineage, constraints, owners, and business rules.

Semantic Drift
The gradual divergence in how terms or data elements are interpreted across systems.

Service Event Object
A governed representation of service actions, tickets, field interactions, used for reliability and churn models.

SHAP (Shapley Additive Explanations)
A mathematically rigorous method for attributing prediction importance to features.

Stillness Index
The measure of operational calm (fewer exceptions, fewer contradictions). At DWL, stillness became a competitive advantage.

System-of-Record Map
The authoritative mapping of which system owns which object (SAP for Orders, Salesforce for Customers, etc.).

T

Tax-Sensitized Data Object
A governed object aligned with financial, tax, jurisdiction, or compliance rules (Margin, Contract, Asset).

Term Integrity
The combination of ownership, lineage, and usage consistency that makes a definition trustworthy.

Truth Arc
The historical pathway of how each function understood performance before alignment — used as a reconciliation tool.

U

Uncertainty Pause Trigger
The condition that forces a model to halt when semantic uncertainty exceeds acceptable thresholds.

V

Variance Signature
The historical pattern of how a customer, plant, or process fluctuates — used to detect anomalies.

Visibility Layer
The reporting, dashboard, and explanation surfaces that reveal aligned truth to humans.

W

Wisdom Audit
A periodic review of whether governance, definitions, and models are being interpreted consistently across teams.

Wisdom Rules
Codified principles used by AI and humans to maintain semantic alignment, detect drift, and ensure ethical reasoning.

Wisdom Threshold
The boundary at which decisions require human judgment rather than machine logic.

Table of Figures

Index